UKRAINE

UKRAINE

The Good, Bad and the Ugly

From the Bucha Massacre,
Liberation of Kherson to the Fall of Bakhmut

PETER SAVAGE

Publication Disclaimer

We have changed the names, some organisations, and the specific sequences of events to protect the people involved in this publication. This book is based on the author's personal experiences and recollections. While every effort has been made to ensure accuracy, certain names, characters, and identifying details have been changed to protect privacy. The author does not accept liability for any loss or damage alleged to arise from the contents of this publication. The identification of any individual in this story is purely coincidental. The author and publisher accept no responsibility or liability for any perceived resemblance to actual persons, living or dead.

Other books in this series, Putin's Pathway

Transforming Hobby Drones for Warfare

Part Time Soldier, Full time Warrior

www.petersavagebooks.com

Acknowledgements

In the course of crafting this story, I have been privileged to delve into the remarkable tapestry of Ukraine's history, culture, and resilience. It is a story marked by moments of triumph, as well as heart-wrenching challenges. None of this would have been possible without the unwavering dedication and support of several individuals and organisations, to whom I owe a profound debt of gratitude.

First and foremost, I extend my deepest appreciation to the brave and resilient people of Ukraine. Your unwavering spirit and determination in the face of adversity have been a constant source of inspiration throughout my journey. Your courage in the most trying of times is a testament to the indomitable human spirit.

I would like to express my heartfelt gratitude to the legitimate aid agencies and overseas volunteers operating in Ukraine. Your tireless efforts to provide humanitarian assistance and relief to those in need have not only saved lives but have also restored hope in the darkest of hours. Your commitment to alleviating suffering and fostering a better future for the people of Ukraine is commendable. UK Aid, DIY Ukraine, Ukraine Patriots Kyiv, Unbroken Lviv, numerous hospitals and children's charities in Kyiv and Lviv and UK Med.

To my friends and family, who provided unwavering support and encouragement throughout the visits and writing process, I am profoundly grateful. Your belief in me and this project has been my driving force.

Last but not least, I want to express my gratitude to the readers who embark on this journey with me. It is my hope that this book sheds light on the remarkable story of Ukraine and its people, and that it serves as a testament to the enduring strength of the human spirit.

Together, we stand in solidarity with Ukraine, and I dedicate this story to the resilient souls who continue to persevere in the face of adversity. With heartfelt appreciation.

Poem, The Waste

War turns cities into echoes and names into smoke.
No victor rises from the rubble, only survivors counting ghosts.

In the end, the battlefield stays hungry and the world grows smaller
for every life it swallows.

Contents

CHAPTER 1

Introduction

Many people ask me why I went to Ukraine during the war. It is a difficult question to answer in many ways, yet it seems very simple to me. As a young man, I had been conditioned in the British Army from day one to see the Russians or, more importantly, Putin's regime as the enemy. For years serving with the Army reserves in REME, (Royal Electrical, Mechanical Engineers) and Signal corps, we were trained diligently in the art of warfare and our trade-craft while unknowingly being exposed to more and more evidence and conditioning of how the enemy operated and repressed their people. At the time, I worked behind the Berlin Wall on six- week detachments, listening to the Russians as part of a listening station helping the East Berlin People's Resistance Party get families out of East Berlin to the West. It was an exciting time and gave me a deep insight into how the world worked and what it meant to live in a free, democratic society. I understand that there were limitations and problems with democracy like anyone else. Still, I would much rather live in a free world than in a subjugated communist empire willing to discard, kill, and imprison its people so readily when they disagree with their government, even though I still understand that people also disappear occasionally in the West.

When the war in Ukraine started, I already had an insight into what the people would face and the indignities they would endure under the Russians. I also shared the euphoria experienced by the families of Berlin when the Berlin Wall was demolished by the people in 1989. I had been there and used the sharpest knife I could find to chisel out a small piece of concrete from the wall to fix the day in my mind forever. After that experience, I felt compelled to try and support Ukraine in any way I could and resist the invasion of their country.

I learnt a lot in Ukraine, and for the first time, I finally comprehend the reasons behind my grandfather's reluctance to discuss the two World Wars. During the First World War, he was drafted into a logistics company due to limited sight in one eye, preventing infantry enlistment. Throughout the war, he diligently served in the trenches of France, orchestrating the movement of equipment across Europe to the front lines. He received his discharge on Armistice Day in 1918. But his dedication didn't stop there; he then volunteered to lead an Anti-Aircraft Battery on Hackney Marshes in London. This was part of the 435th (Mixed) Heavy Anti-Aircraft Air Defence Unit of Britain's Royal Artillery. He spent another remarkable five years defending London during the Second World War and the Blitz, showcasing an astonishing display of endurance and commitment to both his family and England.

Though I had also served previously, I had never encountered the realm of total mechanised warfare. I can now finally fathom the horrors he witnessed and the emotional toll it exacted on him and all those involved. Open discussions about war remain scarce, primarily due to the distressing nature of the topic. The act of revisiting the experiences invokes the resurgence of terror and memories of the torment endured.

Those who haven't personally faced it can never truly grasp our sentiments regarding the harsh reality of war, those intertwined sensations of fear and excitement, coupled with the overwhelming helplessness and sickness upon encountering death and the chaos of total warfare.

In September 2022, I finally contacted the Australian Ukraine consulate in Canberra to find out what I could do as a civilian or registered Chaplain. At my age, being over 50, there seemed to be little I could contribute. When I did finally get through and spoke to someone days later, Volodymyr, who was one of the senior consulate officials at the Australian Ukraine Embassy, he talked to me, emphasising that he could not recommend a visit to Ukraine due to legal limitations in Australian policy. Ukraine was a dangerous place to visit, his hands were tied, but he appreciated that Ukraine was gaining a lot more support from the people of Australia. The only thing he could do was to put me in touch with another man who had previously been to Kyiv and could give me more information.

Shortly after that, a guy from the East Coast called Damo contacted me, and the journey began. Like me, Damo was an ex-serviceman passionately involved with Ukraine and had been there for three months in the summer of 2022, working in the community and teaching first aid to troops just back from the front lines. He was well qualified, having served with the Australian Defence, 2/14th Light Horse Regiment, now totally focused on providing more support for the people of Ukraine and just wanting to reduce their suffering.

I continued to talk to him over the next week as he managed my application, vetting my qualifications and aspects of my service record before he gave me the go-ahead to visit, after which I decided to book flights to Warsaw and possibly on to Krakow before moving to the border to meet a representative of the Ukraine Aid community. It was challenging to get a genuine answer from anyone about how you could help in Ukraine, even the conditions of service within the country or any other aid options. The plan was fluid, to say the least, and changed daily until I left for Poland in mid-October, as I became more aware of other contacts on the ground through the Embassy network. It became clear that a new team was being developed that could connect me with local resources and people on the ground. Above all, I knew that the conditions on the ground would be challenging and I needed to connect with the right people quickly and find a local fixer, someone who knew the local landscape, customs and culture.

To ensure a productive trip, I needed a flexible schedule that could be changed daily. There was so much to take in and plan for it wasn't easy to make any definitive decisions. There was one thing I knew for sure, we would have to wait to get answers in Australia, and I needed to visit Ukraine before I knew whether we could be of any use. After creating an initial plan, I decided to fly into Warsaw before travelling to Lviv and then on to Kyiv to meet up with a guy from a UK Aid- based organisation who had been giving us some information about the situation on the ground finding out how organisations worked without losing part of their consignments at the border or to organised crime gangs posing as embedded charities or NGO's, (Non-Government Organisations).

Poland

By the time I arrived at the airport, I was over excited having spent long days preparing for the trip gathering the right equipment, money and supplies together with desperately trying to compile a simple list of essential words and phrases to translate making the trip more practical. The flights from Australia to Poland were long, tiring, and laborious, spanning a vast distance and multiple time zones; the journey required me to brace myself for a marathon of travel. The first leg of the journey was the worst, a lengthy flight from Australia to Qatar. This flight alone could stretch for more than 12 hours, subjecting passengers to cramped seats, jet lag, and limited mobility. I spent hours waiting in bustling airports, navigating immigration, and enduring security checks, testing my patience even as a seasoned traveller. The anticipation of reaching the final destination in Poland and starting the task kept me going. While the flights from Australia to Poland were undoubtedly exhausting, the reward of experiencing a new country and culture made it a journey worth enduring.

On arrival in Poland in October 2022, I took a short taxi ride to the Hampton Hotel in the city centre. Eager to test my newly learned Polish, I quickly realised how inadequate it was judging by the puzzled look on the drivers face as I stumbled thorough the address. I was keen to start and quickly settled into a room on the seventh floor, immediately booking a coach online from the local railway station in Warsaw to Lviv. After such a long f light, I sat in the bar and had a beer to decompress, talking to the bar staff, who spoke fluent English. They were surprised to see an Australian; it seemed novel to them, particularly when I had travelled so far to take an interest in the conflict. Perth in Australia is the most isolated city in the world and no one in the hotel even knew it existed. I quickly began to understand that many considered they were also at war with Russia. They feared the Russians and believed that NATO would desert them as the war progressed. We discussed the conflict for a couple of hours. I tried to reassure them that NATO was already involved in the war as troops were stationed in Poland and supported Polish troops training Ukrainian personnel. Understandably they were still terrified as Poland shared a border with Belarus who supported Russia in the war; in fact, most of the younger people in the hotel felt scared that they would soon be invaded once Russia formally engaged NATO. To them, it only seemed a question of time before a world war or invasion of Poland. I went to bed that night, uneasy with the conversation, feeling worried that the whole situation was close to escalating into a global conflict. I was so tired and unsettled with the thought of all-out war was horrific; I returned to my room and fell into a short restless slumber.

Warsaw, City Centre, from the Centrum Hotel

The next available coach was just hours away; I had a quick shower and a protein bar before leaving the hotel just after midnight. Straight away, there was a real issue with the language barrier; not speaking any Polish would be a big problem. I had yet to find out where the coach was and if it had arrived. I nervously wandered around, trying to find someone who spoke English. Ironically the first person who spoke some English was a young Russian girl willing to help me find the coach. She felt lonely being Russian and ignored by many other passengers. The people were friendly enough when they realised I was from Australia; however, they were all uncomfortable with the thought of travelling back into Ukraine as most were unsure about what they would discover when they got home. Many of them had escaped Ukraine at the beginning of the full-scale invasion in February 2022 and just wanted to go home, sadly many of them had lost relatives during the first part of the war during the spring. Most of them were refugees. The gravity of the situation began to sink in as I looked at the people around me. Most of them had lost everything, including the men in their families who had been conscripted or already injured or even killed in action. They were a mixed group of middle-aged women, children, and older men, agitated, tired, with desperation etched into their faces.

CHAPTER 2

The Border

I settled into my seat on the coach and tried to get some sleep. It was a six-hour drive to the border, and I was a little nervous myself. I just had no idea what to expect. We had one comfort stop on the journey at a service station that was limited to 10 minutes and quickly ushered back onto the bus without compassion. When we finally got to the border crossing, there was a long queue of traffic that stretched for miles. The coach sat for ninety minutes quietly edging forward before, we even saw a border guard and handed in our passports. We were all taken off the coach and escorted through the Polish passport control point with our bags towards the Ukraine border.

As the passengers approached the Ukrainian side of the border and their soldiers came into view, everyone stopped talking as a solemn silence descended across the group.

In most cases, passengers were interrogated, and their bags tipped out and emptied on to a counter for inspection. Surprisingly when it came to my turn, my passport was returned to me immediately stamped, and the guard waved me through without another word. I was curious to know whether the guard was just being supportive or could not speak English. However, after a little thought I realised the same had happened to me at the airport. I began to understand that the Polish and Ukrainians were thankful for our support and wanted to help us in return in any way they could relinquishing some of the formalities.

Ukrainian, Poland Border

The border guards were more interested in people from non-EU countries getting into the country and any Ukrainians escaping conscription. Within our group, there were a few Asian people and other non- English speaking people within the group. The guards were more interested in these people and wanted to ensure that they were genuine visitors. The guards emptied all their bags searching them in depth whilst being interrogated. A large senior guard removed a couple of younger guys travelling together with a middle eastern appearance from the coach, and we didn't see them again.

From a humanitarian point of view, the process and experience had been distressing. The passengers had been unsettled to start with before they had been sorted through and degraded. It had been a disappointing necessary evil. To make matters worse, we had to go through the process twice, first on the Polish border, then on the Ukrainian border after crossing no-man's land. Overall, the experience took four painful hours to complete. While the passengers were filtered through reasonably quickly, the search of the bus took hours. At first, the guards searched it from top to bottom, then to my surprise just as I thought it was over, the vehicle was taken into a large hanger and practically stripped and searched again with sniffer dogs.

Rava-Rus'ka

As soon as we entered Ukraine, the attitude of the people around me changed to an upbeat mood. Understandably, many of them were pleased to be home. There was a stark difference in the landscape and architecture. Outside the window, the countryside unfolded in a completely different character: wide fields framed by forests, weathered cottages scattered among the hills, and church spires gleaming in the pale morning light.

Our first pause came at Rava-Ruska, where the driver announced a brief stop for coffee and rest. The town rested gently among green slopes and orchards, its calm presence a welcome contrast to the long journey. Buildings showed a blend of regional design and spiritual influence , a tapestry of folk artistry and ecclesiastical form. Winding lanes of uneven cobblestone meandered between cottages painted in soft hues of blue, rose, and cream. Wooden fretwork framed their doors and windows, each pattern distinct, each telling a local story through its carving.

At the centre lay the rynok , the market square , alive with movement and sound. Arched walkways lined the edges, sheltering traders who arranged baskets of fruit, embroidered linens, and hand-forged trinkets. The façades around the square displayed delicate plaster flourishes, showing both the elegance of the Baroque and the restraint of the Neoclassical. From the balconies above, trimmed in ornate timber lattice, residents watched the bustle below, leaning on rails polished smooth by generations.

Rava-Rus'ka Orthodox church, in town centre

Standing sentinel over the town was the Orthodox church, a striking monument of faith and craftsmanship. Its bulbous domes, plated in gold, glowed against the sky, and the tall crosses crowning them caught the morning sun in flashes of brilliance. The structure merged the solidity of stone with the warmth of wood. Along its outer walls, carved reliefs portrayed harvest scenes, flocks, and celestial emblems, the handiwork of master carvers whose devotion lived on in every chisel mark.

Inside the church, silence seemed to breathe through colour and light. Beneath a canopy of frescoes and saints' portraits, candles flickered in brass holders, their glow reflecting on gold-leaf halos. The faint perfume of myrrh and beeswax drifted through the

stillness. The altar was softly illuminated by sunbeams filtered through stained glass, scattering pools of amber and blue across the floor. A magnificent iconostasis screen divided the sanctuary, its painted figures seeming almost alive in the candlelight, inviting contemplation and prayer.

Beyond the square, smaller lanes wound away toward quieter quarters. Modest homes with steep roofs and overflowing flower boxes stood shoulder to shoulder. Many had verandas where neighbours gathered, sipping tea or simply observing life unfold. The atmosphere was one of ease, the rhythm of the place untouched by haste.

At noon, the bells from the church tower rang out, their notes rolling through the air like waves of song. The town seemed to pause, a collective breath, a moment of reverence. Here, faith and artistry existed not as opposites but as partners, shaping a setting that felt both sacred and alive.

For me, it was a glimpse of the Ukraine that should be , peaceful, industrious, and bound to its traditions. The tranquillity of that small community stood in sharp contrast to the knowledge that the wider country was still enduring conflict.

We boarded the coach again at a given time to continue the journey and settled back into the trip. Most of the people on the coach ignored me with the general disdain associated with being a foreigner abroad, unable to understand or speak their language. However, some English-speaking locals were genuinely interested in who I was and why I was travelling to Ukraine. Whilst I was keen to interact with the people around me, I gave them only a general overview of the trip, avoiding specific details, sharing too much would have been irresponsible. After the border crossing, however, most of the passengers on the coach had became noticeably more interested in me, having seen the way in which the guards had treated me, ushering me through without question or delay.

By the time we reached Lviv, exhaustion had replaced curiosity. As soon as the coach halted, everyone hurried to disembark. I gathered my bag and slipped away into the moving throng, careful to draw no attention. I had already become the subject of too much interest or even worse suspicion.

CHAPTER 3

Lviv

Lviv, one of the advantages of talking to Damo and the Embassy was that he could provide me with a list of hostels and reasonably priced hotels. I used one of his suggestions and booked into the Dream Hostel in Lviv, just off the city centre, a perfect location down several cobbled streets. Lviv was beautiful, a historical city full of architectural monuments dating back to the 14th-17th centuries, and I could see why UNESCO had included it in the World Heritage Listing. Stunning, it was hard to believe the city was under siege; the people looked happy and at ease. The only problem I had was the cobbled streets; there were so many that it became difficult to pull my suitcase across the pavements. The hostel was basic but clean and comfortable, with shared rooms and a kitchen. I found a lady at the hostel who spoke English and she was delighted to help me find a mobile phone shop just outside the hostel to buy a new phone sim connecting me to the network. Poland had its phone network integrated with the European network where most phones worked, but Ukraine did not offer roaming or a connection to an established EU network, so it was challenging to communicate with anyone at home or use Google Maps. Even though I managed to get a sim installed and could make calls, the connection was patchy, and most of the apps on the phone were limited. I could call, but the apps could not send and receive messages reliably. I spent the next few days in Lviv, acclimatising to the new environment and began the job of connecting to some of the new contacts in our group that were provided by the Embassy.

Most of the people in the hostel were friendly and kind enough to help out an interested aid worker from Australia, being patient with me not knowing any of the local languages or dialects. There were a mixture of students, other aid or support workers, international businessmen and the press. I was amazed with the mixture of how many international people were there and surprised with the amount of UK, American and Australian workers that I was coming across. There were hundreds of young Americans and Europeans that I had already met who wanted to support Ukraine, some of whom had never been in the military and were deeply disappointed that they had been turned away from the Ukraine Legion and had subsequently signed up independently with a Battalion separately or volunteered with an Aid agency.

It became apparent that no one was being turned away, everyone who wanted to volunteer or help could find a place to work regardless of their age or background. The more qualified you were, the easier it was to find somewhere to go. In contrast to what we had been told over the phone in Australia, once you were there on the ground, any organisation or military unit would welcome you in to their ranks. The more I listened to people's stories, the more I understood that just having another pair of hands to help was enough.

The hostel's proximity to the city centre provided ample opportunities for exploration, with numerous shops to gather supplies. I wandered through the back streets of Lviv, making an effort not to appear out of place, trying to blend in, while the city centre buzzed with new experiences, ranging from live music to outdoor dining spaces and public art. It resembled any other vibrant and exciting European city until the air raid sirens suddenly sounded, and everything changed.

I had received a list of people from Damo with which to engage with and continue to build our network. Among these contacts was a local official introduced to me through the Ukraine Australian Embassy, whom we referred to as "Vasily", a code-name to safeguard his identity. I reached out to him, organised a meeting to discuss opportunities for aid in Ukraine as well as priority areas requiring volunteers. Capacity building was going to be essential if Ukraine was going to survive a protracted war.

The First Air Raid

The First Air Raid. My second day in Lviv started with a late breakfast. Whilst the hostel offered lunch and evening meals through a separate restaurant, their breakfasts were self-catering. As I hadn't had an opportunity to go shopping yet, I walked around the cobbled streets of Lviv for ten minutes looking for a café or sandwich bar. A small café designed around the theme of a bumble bee caught my eye, so I went in and asked for a menu. Almost all the menus I came across were available through a bar-code on the table read from your mobile phone. As there was little to no internet available, I spent a couple of minutes trying to decipher what was available with the manager in broken English. I ordered what I thought was a simple omelette and coffee which turned out to be a gourmet option based on brie and eggs. It was great, so I tucked in. Only minutes after I had started breakfast, just as the coffee appeared, the air raid siren went off; its synchronised sound came from my mobile phone app and echoed outside from claxons across the city. The younger teenagers in the café looked terrified, glancing at each other trying to figure out what to do before momentarily disappearing into what I assumed was the cellar air raid shelter. I sat for a few seconds looking at the café manager, who just shrugged and continued to read the paper; I figured out what the hell and decided to eat my breakfast rather than facing the possibility of being buried alive in the cellar if we had been hit.

The Bumble Bee Cafe in Lviv

After another minute or two of sirens, it went quiet outside as you could feel the anticipation of the approaching missiles and drones. The information on the phone app gave us a time to impact, so the manager and I chatted over breakfast and shared a coffee. The omelette was the best meal I had eaten in days, and I didn't want to walk away from it, letting it get cold, regardless of how sick the approaching missiles made me feel. There were a few distant impacts and explosions as the missiles hit home, with one impact closer than I had anticipated making me jump and almost spilling my drink. I finished my meal and contemplated the experience of having survived my first air raid in Ukraine. We sat quietly for a while, finishing our drinks before we heard the all-clear siren sound just before the sound of the ambulances and fire engines could be heard in the distance, racing towards the destruction just down the street. This was the first time I began to understand the gravity of my situation. I left after breakfast nervously to continue my exploration of Lviv and the local area.

As I left the restaurant, I saw that a missile had struck the city just streets away. Standing in the doorway beside the manager, I shook his hand in farewell, our gaze drifting across the road toward a rising column of smoke. The acrid scent of burning plastic and rubber clung in the air making me wretch slightly before I covered my face with a scarf to help mask the toxin.

I stood there, transfixed, struggling to comprehend that such destruction had occurred so close and indiscriminately. Powerless to help, I prayed that no one had been harmed, though in my heart I knew it was likely that people had been injured, or worse.

After a time, I realised I had been lost in the moment, overwhelmed by the scene before me and the manager had already walked back inside. Forcing myself to break free from the trance, I began to think of the tasks ahead and the meetings planned for the afternoon, turning away and beginning the slow walk back to my accommodation trying not to dwell on the carnage that had been left behind.

CHAPTER 4

The CIA Intelligence Contact

The new contact in our group was from Odessa, Vasily had recently joined to help build capacity and advise us. I was keen to meet him having received the task to engage with him while in Ukraine. So, I scheduled a lunch meeting with him in a quiet cafeteria off the main town square.

Vasily had agreed to wear a bright blue cap carrying a briefcase while standing in the doorway of the Opera House, just off the central square. He was easy to spot. What unsettled me more was the fact that he had already picked me out from over five hundred metres away as I crossed the square. I must have looked completely out of place. He greeted me with a broad smile and a firm handshake before we set off towards the nearby restaurant he had recommended. As we strolled across the street, I joked that I must have looked like a tourist. He chuckled, but then his tone quickly shifted to a more serious one as we agreed that I needed to sharpen my trade-craft and find a way to blend in more inconspicuously.

He was right, of course. I needed to present myself as someone professional, prepared for the challenges ahead, not a clueless outsider. I'd already been warned; looking like a war tourist was dangerous. The locals had no patience for that kind of naivety. This wasn't some adventure, it was serious business, and I needed to meet him on those terms. Once we had sat down to order lunch, I began by asking him for advice and ordered the national dish of Ukraine, Borscht a well known beet-root based soup.

During our conversation, we addressed the challenges of delivering aid to the country amidst the ongoing war. These challenges stemmed from outdated procedures, corruption, and logistical hurdles. Vasily and his associates were individuals genuinely dedicated to supporting Ukraine. They aimed to enhance the flow of aid into the country and establish reliable connections with authentic aid organisations and contacts.

I quickly recognised that Vasily and his two companions who had just joined us were government employees from Ukraine and had been briefed about my visit through the Embassy network. My purpose in meeting them was to receive guidance during my visit and ensure my safety. We all recognised the necessity of connecting the Australian Ukrainian community with aid agencies in Ukraine while sidestepping local political restrictions, corruption, and biases from the Australian community.

The meeting went well. We were open and honest with one another, united by a common goal: strengthening support for Ukraine from Australia and the Commonwealth. After an hour, Vasily acknowledged that I had answered many of his difficult questions realistically and in a way that met his expectations. This was a positive development, as trust between us was essential moving forward. He needed to understand not only why I had volunteered for this work but also that I was capable of carrying it out with honesty and professionalism, without placing myself or others at risk.

Vasily, was a knowledgeable individual who had a background in the Ukrainian Defence Force and connections to the CIA, he provided me with real-time insights into the situation in Ukraine. He conveyed the intricate reality and the political dynamics within the country. While Ukraine aimed for freedom and integration with Europe, Vasily was candid in noting the desires of certain segments in the Eastern part of the country and the Donbas province who wished to maintain their ties to Russia.

The complication arose from these groups being in the minority and having been influenced by the Putin regime to establish counter-terrorism units in these regions. Over decades, Russia had been relocating patriotic Russian families and sleeper agents to these areas, undermining Kyiv's authority and preparing the region for conflict. This intricate issue was further complicated by the fact that many families in Ukraine had relatives and descendants in Russia. The majority of these people sought peace, not war.

We all agreed that Putin's long standing objective had been to regain Crimea as a strategic peninsula, directly connecting it to Russia. This objective had triggered the war in 2014, and since then, an all-encompassing conflict seemed inevitable.

By now my head was spinning with the implications of the conversation, as I was beginning to understand why it had been so difficult to obtain a genuine answer before my visit. The truth about the situation on the ground was both unpleasant to hear and distressing to comprehend. It was a deeply complicated reality, entangled with organised crime gangs flooding in from Europe and preying on the most vulnerable within the population. Hundreds, perhaps even thousands of women and children had disappeared through human trafficking, while countless aid deliveries had simply vanished. Most alarming of all, elements of these criminal networks had begun infiltrating both government institutions and the Army's command structure, bolstered by Russian financing aimed at destabilising Ukraine's efforts to support its own people. This, I realised, was the stark reality of war.

The importance of allied support for Ukraine had become unmistakably clear. The United Kingdom and Europe stood firm, but most critically, Ukraine depended on the continued backing of the

United States and its intelligence community. American involvement was not abstract or symbolic; it was operationally decisive. Access to real-time satellite imagery, signals intelligence, and rapidly shared battlefield assessments shaped everything from front-line manoeuvres to humanitarian logistics. The flow of live intelligence, whether from orbital platforms, technical surveillance, or trusted contacts on the ground like Vasily, provided the situational awareness necessary to navigate a fluid and lethal environment. For those of us coordinating large-scale aid deliveries from overseas, that information was not a luxury; it was the difference between efficiency and disaster. Convoys could be rerouted around emerging threats, distribution hubs adjusted to avoid missile risk, and supplies directed precisely where need was greatest. Simultaneously, military planners relied on the same intelligence picture to anticipate strikes, reinforce vulnerable sectors, and integrate air defence responses.

In modern conflict, information dominance underpins survival. Without the timely sharing of actionable intelligence among allies, both humanitarian operations and defensive strategy would have been dangerously blind.

After lunch, we agreed to connect with the Australian team and Embassy representatives to review our progress before deciding on the next steps, though I already had a fairly clear idea in my own mind of the direction I wanted to take. With this in mind, we returned to the hostel for a quick online Zoom discussion and catch-

up.

Speaking again with the team in Australia and weighing my options, it became clear that I needed to gain a deeper understanding of how the aid community functioned within Ukraine. This knowledge was essential before we could determine the best way to move humanitarian assistance across the border and ensure it was distributed fairly. Although there was no shortage of support available from Australia, particularly from the Australian Federation of Ukrainian Organisations (AFUO), almost everyone I spoke with had their own ideas about how best to achieve this. However, many lacked an appreciation of how aid agencies operated on the ground in different regions, or of the extent to which distribution was influenced by local connections and political considerations.

It was then that I realised I needed to spend time with several local aid organisations to observe how they worked, understand what shaped their priorities, and learn what truly influenced the flow of aid. To gain real insight, I had to become an insider and confront the reality of war and international aid on a large scale.

From my time with the British Army, I had already begun to grasp the cultural differences and the unique dynamics of dealing with those in positions of power. Yet Ukraine was unlike modern Europe. In many areas, local politicians had assumed roles as battalion commanders, using these positions to preserve both their influence and their income. While the backbone of the Ukrainian Army consisted of professionally trained officers and an elite, battle-hardened core, many of the other battalions were raised directly from the commanders' villages, towns, or regions.

What I did not initially grasp was that most aid agencies and NGOs were deeply embedded within the military for practical reasons, enabling them to function more efficiently. Even well-established international organisations such as the Red Cross and UK-Med were closely associated with the military. They employed "fixers", local Ukrainians who could speak multiple languages, understood the terrain, and, perhaps most importantly, navigated the cultural and social fabric that shaped how things truly worked. Connecting with an agency and "fixer" would be essential to understand how everything worked.

I reached out to Andy from UK Aid, whose name had appeared in our social media group, and arranged to meet him, volunteering for a couple of weeks to join his team. I discussed the idea with Vasily, who was already familiar with UK Aid and agreed it was a good opportunity.

That afternoon, after lunch, we sat quietly while he used his mobile phone to book my overnight train from Lviv to Kyiv through "Proizd.au." Vasily even insisted on accompanying me to the station later that evening, making sure I reached the correct platform before boarding. This had been my first significant meeting in Ukraine, a small but meaningful step forward. I felt quietly pleased with myself, and at ease in Vasily's company. He was a warm, sincere family man, with a couple of close friends who shared his easy manner.

We had just paid the bill and were about to leave the alfresco dining area when the distant sound of a brass band began to drift across the square. At first, I thought little of it, perhaps a festival or community gathering, but then Vasily and his friends suddenly rose to their feet. Without a word, they turned toward the church and placed their right hands solemnly across their chests.

The mood had shifted instantly. The music, slow and heavy, carried across the cobblestones as the church doors opened. A coffin emerged, borne on the shoulders of grieving family members, flanked by a military honour guard carrying banners that drooped in the still air.

My chest tightened as I understood what was unfolding. Another Ukrainian had fallen at the front. A photograph of the young soldier was carried ahead of the coffin, his face heart breakingly youthful, barely into his twenties. The grief of his family was raw and unrestrained, their sobs carrying through the silence of the crowd. Soldiers marched in step beside them, heads bowed, their boots striking the pavement in slow, deliberate rhythm in time with the music.

The procession passed just metres from where we stood. The hearse waited at the edge of the square, its engine low and restless, before the coffin was lifted inside. With a slow, mournful roll, the vehicle pulled away, leaving behind only the echoes of the band and the quiet weeping of the family.

For a long moment, no one spoke. The weight of loss lingered in the air. Then, as the crowd began to disperse, I caught Vasily's eye. He gave me a small, gentle smile, an attempt to ease the heaviness of the moment, to remind me of purpose rather than despair.

It struck me then with sobering clarity: this was not an abstract war spoken about in news reports; it was lived, felt, and mourned every single day. The funeral was a stark reminder of why I had come to Ukraine, and it steeled my resolve to do whatever I could to make a difference.

We went back to the hostel to load my bags and gather equipment before loading it in to his car driving to the station. While I expected an increased level of security, I was surprised by the scale of the operation, there were hundreds of soldiers and security guards around the station checking travel permits and identification with an heightened level of security similar to an airport including scanners and body searches. At this point I began to appreciate the value of the Ukraine Embassies letter of association as it provided me with a valuable asset enabling me to pass through most of the security checks unhindered, to help people understand it had been written simply on headed paper in both Ukrainian and English, one side for each language. Simple and very effective.

Vasily was kind enough to come in with me to check that we had bought the right ticket and showed me the direction to the proper platform before we shook hands and reinforced that we would stay in touch.

With a few hours before my train arrived, I sat in a café to decompress, taking in the surroundings and savouring the moment. I had always wanted to experience the culture of Ukraine, and this was the perfect opportunity to absorb the atmosphere. The station was a masterpiece of early 20th-century Art Nouveau architecture, exuding grandeur and refinement both inside and out. Inside, the station revealed an opulent interior: walls and ceilings richly adorned with ornate stucco, polished marble panels, and decorative ceramic tiles. These materials created a striking visual harmony, enhanced by intricate ceiling mouldings and soft, natural light filtering through stained-glass windows.

The waiting halls, arranged by class, further reflected this lavish attention to detail. The first-class areas evoked the elegance of Viennese clubs, adorned with carved wood and marble accents, while sweeping tiled passageways and ornate wrought-iron railings guided travellers through what felt like a palace of transport. The station complex itself was vast and awe-inspiring, easily as impressive, if not more opulent than many of Europe's grandest constructions. As I paused to rest before boarding my train, the experience was everything I had hoped for, offering a quiet moment of reflection before continuing my journey onward to Kyiv.

Train to Kyiv

The train arrived right on time. With the help of a station attendant who pointed me towards the sleeper car, I handed my ticket to the conductor and was shown into a bunk room fitted with four berths. Trains in Ukraine, compared to their European counterparts, felt like hardy, well-built workhorses; overpowered engines of steel designed less for elegance and more for endurance, capable of withstanding long distances, harsh climates, and heavy loads with unwavering reliability. Though the furnishings were simple, the carriage was comfortable, attended by a warm and attentive steward who clearly wanted to make the journey as easy as possible for us. I shared the compartment with a family of three, an elderly man, his daughter, and his grandson. They seemed just as curious about me as I was about them, which made it easy to strike up a conversation with the translator on my phone.

As I settled into my bunk on the overnight train from Lviv to Kyiv, I couldn't help but feel a sense of excitement mixed with a hint of anxiety. The rhythmic clattering of the wheels on the tracks, the gentle swaying of the carriage, and the dimly lit interior created a soothing atmosphere, perfect for a night of reflection and anticipation. The air was stuffy and full of humidity.

I gazed out the window as the train pulled away from Lviv station, watching the city's lights slowly fade into the distance. The landscape unfolded in shifting scenes, serene stretches of countryside, quiet villages, and, at times, moonlit fields broken by sidings stacked with supplies and armoured vehicles. It was in these fleeting glimpses that the war made its presence felt. Even the windows served as a reminder; each was covered with a shatterproof film, designed to withstand the force of an explosion and reduce the risk of glass splintering into the carriage.

The darkness outside enveloped the train, the window blast-proof film making if difficult the see through as the windows began to mist. Four bunk beds in the compartment were shared with the family, a comfortable and cosy sleeper car. The soft, upholstered seats had been converted into snug beds that promised a peaceful night's rest. I exchanged polite nods with my fellow passengers, acknowledging the unspoken camaraderie that developed during these shared overnight journeys.

Soon, the soothing lullaby of the train and the gentle swaying motion began to work its magic. My eyelids grew heavy, and I decided to retire to my bunk having stowed all of my baggage toward the foot on the bed. The warm blankets and the rhythmic clatter of the wheels provided the perfect lullaby, and I drifted into a short, peaceful slumber.

Somewhere in the middle of the night, I stirred awake to the crackle of an announcement over the tannoy. The train kept moving, though at a slower pace, and from beyond my compartment I caught the faint hum of voices drifting through the corridor.

An English-speaking passenger informed me about a one-hour delay in a siding due to air raids along the route. At this stage of the war, most people had been surprised that the rail infrastructure had not been bombed. However, we all understood that Russia desired to control the network intact for themselves and did not want to destroy them.

People now stood in the doors and hallways, gazing east into the night sky as streaks of light arced overhead. Somewhere in the distance, the horizon ignited, another air raid moving toward Kyiv. Trails of fire, anti-aircraft bursts, and distant explosions lit the darkness, a grim spectacle reminding us that life on the tracks, like the war itself, never slept. For over an hour we watched the display, caught between awe and dread, before the train finally began to roll forward once more.

The mood on board the train had shifted. Where earlier many had been cheerful and quick to smile during our first conversations, a heavy silence of dark contemplation now lingered. I needed no knowledge of the language to see it; the mothers softly reassuring their children as little hands clung tightly with fear in their eyes, whispering words of comfort only half-believed. When they looked toward me, I tried to return a gentle smile, as if to reassure them that everything would be all right once they reached home. Yet, from the fragments of conversation I had shared with a few passengers, I already knew the truth: many had endured loss, while others carried the weight of relatives maimed by shellfire at the front. They were returning not out of hope but necessity, with nowhere else left to go. Still, they placed their faith that Kyiv would endure the coming winter under President Zelensky's steadfast leadership.

As we drew closer to the city, it was time to prepare for disembarkation. My thoughts turned unexpectedly to my grandfather and his service on the anti-aircraft batteries in London during the Blitz. I wondered how similar the sight before me was to what he had endured night after night. The thought unsettled me at first, but then steadied me; if he could endure five long years of unrelenting bombardment, surely I could withstand a few months of air raids in Kyiv. Difficult as it was, I forced myself to dismiss my worries as childish, even as I braced for the reality of living under the constant shadow of war.

After nine hours, I knew that our journey was ending. The train gradually slowed, and the anticipation of reaching Kyiv grew stronger. I folded away my bedding, gathered my belongings, preparing to disembark.

The overnight train from Lviv to Kyiv had carried me through the darkness, across the vast Ukrainian landscape, and into a new day. As the wheels ground to a halt, I stepped onto the platform with a mixture of accomplishment and unease, ready to face whatever awaited me in the capital. Kyiv lay before me, vibrant even in the depth of night, a city both daunting and full of promise.

CHAPTER 5

Kyiv

I had reached Kyiv in the early hours, making my first mistake of the trip by arriving at two in the morning when the cold was at its sharpest. Even with thermal clothing, the chill cut through me. I should have anticipated that such an hour would bring more than just discomfort.

The station's dimly lit underground tunnels felt foreboding, and almost immediately four drunken Ukrainian conscripts in uniform singled me out. Herding me into a shadowed corner, they made it clear they intended to rummage through my luggage, more out of opportunism than duty. Young, inexperienced, and emboldened by drink, they began emptying my backpack onto the floor.

For several tense minutes, I tried to reason with them, struggling to contain my frustration. At last, my patience snapped, and I barked "Diplomatico!" The word startled them into hesitation, their eyes flicking uneasily between one another. Seizing the moment, I produced the Embassy letter of association showing it to them. They muttered "Diplomatico" under their breath, their expressions shifting from surprise to unease before they reluctantly stepped back.

I fixed them with a steady gaze, waved them off, and began repacking my belongings with deliberate calm. They were drunk, young, and naïve, and I was fortunate their greed had not gone further. The encounter left me shaken, but I understood such moments would become an unavoidable part of the journey. Travelling alone made me vulnerable, and I would need either to steel myself for more incidents like this or seek the safety of companionship. I knew I had to project confidence and authority, enough presence to reaffirm my position when tested.

Slinging the pack back onto my shoulder, I moved on through the station, already thinking about my next steps.

Kyiv station rose before me as another testament to the grandeur of Ukrainian architecture, a vast expanse of echoing halls, gleaming marble, and tiled walls that seemed to stretch endlessly. For a moment I wanted to linger and take it all in, but the circumstances left me little time for admiration. I now faced a choice: remain in the relative warmth and safety of the station until morning, or press on into the uncertain darkness beyond its doors.

Outside, the station steps opened into a near-empty street where only a couple of taxis idled in the cold, engines rumbling softly in the silence. The drivers lingered with an air of indifference, though it was clear they knew they held the upper hand at this hour. I showed one of them the address written on a scrap of paper, and after a long pause he nodded, naming a price so inflated it bordered on extortion. Yet with no other options in sight, I accepted, aware that in that moment I was at the mercy of strangers.

After a short drive through the city centre, the taxi dropped me at the entrance to the hostel, tucked away down a narrow cobblestone side street. I had arrived with confidence, reassured by the staff in Lviv that a night porter would be on duty to admit late arrivals. Reality, however, proved otherwise. My knocks went unanswered; the reception was dark, and no one came to the door.

Left standing in the cold courtyard, I waited, the night air cutting deeper with every passing minute. The temperature hovered just below freezing, and it quickly became clear I would have to endure until morning. Thinking desperately of a solution, I scoured the courtyard and discovered a patio heater tucked beneath an awning. With some improvised tinkering, I managed to hot-wire it into life, its faint glow offering a measure of relief from the biting chill.

Huddled beneath the heater, wrapped in a survival blanket, I fought off exhaustion with whispered songs and mental sing-alongs, anything to stay awake and ward off the numbing cold until dawn.

When someone finally came to the door at five thirty, I was desperately keen to get inside to thaw; even then, I had to wait until breakfast before they would book me in and let me settle. By the time the day was underway, I was relieved to be alive, albeit hungry and keen to explore.

Kyiv, Saint Sophia Cathedral

It didn't take long before I sensed I would feel at home in the hostel. As I waited in the lobby for the receptionist to appear, the morning unfolded around me in a gentle rhythm. Guests emerged from their rooms one by one, drawn by the smell of freshly brewed coffee and the faint hum of chatter from the common area. There was a warmth to the place that cut through the October chis, a sense of easy camaraderie among strangers bound by circumstance. Within minutes, I found myself exchanging smiles and conversation with people who would, in time, become far more than casual acquaintances.

Among them was a young man named Max. His uniform caught my eye before his accent did, and when he heard mine, his expression lit with recognition. "Australian?" he asked, the word carrying both curiosity and friendliness. I nodded, and he gestured for me to join him at his table. Over a simple breakfast of coffee and bread, we spoke easily, as though old friends reunited in a foreign city.

Max was English by citizenship, though his family story was a tapestry of Eastern Europe, Polish-Ukrainian parents and a Russian grandmother. That blend of identities, once a quiet family trait, had become something heavier since the war began. He told me of the regiment he had recently joined, his voice steady but shadowed by the weight of what lay ahead. There was pride in his words, but also a quiet acceptance, the understanding that soldiers rarely choose the shape of history, only the small corner of it they occupy.

The Dream Hostel, tucked away in Kyiv's vibrant Podil district near the Dnipro River, was modest yet alive with energy. Its narrow corridors echoed with the footsteps of travellers, aid workers, and volunteers who passed through on their way to the front or to the eastern border. It was perfectly placed, close enough to the government quarter to feel connected, yet far enough to remain cocooned in its own rhythm.

At that time, none of us had a clear idea of how many Australians, Kiwis, Europeans, or Americans were serving in Ukraine. Information was fragmentary, passed through rumours and second-hand stories. Yet even in those early days, I had already met two Australian Army veterans and several Americans. The foreign presence was unmistakable, hundreds of men and women drawn from distant nations, bound not by nationality but by conviction, curiosity, or something in between.

Through Max, I learned how many of the younger foreign volunteers had been absorbed directly into Ukrainian units, bypassing the International Legion altogether. The Legion itself was vast, thousands strong, but its structure was chaotic, and many preferred the autonomy of the regular battalions. These were not mercenaries in the traditional sense; most were idealists, drifters, or ex-soldiers seeking purpose. Max spoke of them with both admiration and caution. "Some come for the cause," he said quietly, "and some come for the story."

Later that morning, I set out on foot to explore Kyiv. The moment I stepped outside, the city unfolded around me like a living museum, elegant, bruised, and defiant. Its skyline rose with domes and spires that gleamed in the low autumn sun, each one telling a fragment of a story that stretched back a millennium. The air was crisp, and the faint sound of church bells drifted from somewhere beyond the river.

The golden domes of Saint Sophia Cathedral caught the light first, dazzling against the pale sky. From nearly every corner of the city, they shimmered like a promise. Built in the 11th century, the cathedral stood as Kyiv's spiritual heart, its mosaics and frescoes a testament to centuries of devotion. Inside, the air was still and cool, scented faintly with candle wax and incense. I lingered before the icons, struck by the patience of their creation, tiny fragments of glass and stone assembled into images that had outlasted empires.

Saint Sophia Cathedral, Kyiv

Not far away, the Kyiv Pechersk Lavra sprawled across the riverbank, a labyrinth of monasteries and churches, their gilded domes glinting above the Dnipro. The sound of distant bells carried on the wind. Standing there, it was impossible not to feel small. The Lavra had survived Mongol invasions, Soviet desecration, and now, the threat of missiles. It was not just architecture; it was endurance made visible.

Further into the city stood the National Opera House, its grand facade rising proudly from Volodymyrska Street. Built in the late 19th century, its Neo–Renaissance columns and marble halls evoked an age when beauty itself was a form of defiance. Kyiv, I realised, was not a city frozen in time. It was a place where the old and new coexisted in fragile harmony. Sleek glass towers , the Gulliver and Parus business centres, mirrored the domes of ancient cathedrals, symbols of a people determined to rebuild even as the war pressed close.

Eventually, my walk brought me to Maidan Diagnostically, Independence Square. The plaza, once a theatre of revolution, now stood transformed into an open-air memorial. Burnt-out Russian tanks and armoured vehicles were displayed across the paving stones where fountains once danced. Their twisted metal forms, scorched and silent, told a story more eloquent than any monument could. Children climbed cautiously onto the hulks as their parents stood back, reading plaques and taking photographs.

Burnt out Russian Amour, Mykhailivska Square, Kyiv

One inscription bore the now-famous words of President Volodymyr Zelensky: "I don't need a ride; I need ammunition." The phrase hung in the cold air like a vow. Around it, the city's pulse seemed to quicken, a collective reminder that Kyiv had stood when the world expected it to fall.

Amid the wreckage, one object drew my attention more than any tank or troop carrier: a small green family saloon car, riddled with bullet holes. It sat alone, cordoned off but unguarded, its glass shattered and paint scorched. The story attached to it was almost unbearable. A family of four, two adults and two young children, had been trying to flee Bucha, white flags tied to the windows in a desperate plea for mercy. Russian soldiers opened fire anyway. When the shooting stopped, the car was searched, the bodies pulled out and burned to erase the evidence. The vehicle had been brought here as proof, not of victory, but of truth. I stood before it for a long time, unable to look away. The image of that car, and what it represented, would haunt me for months.

Small green family saloon riddled with bullet holes

As the afternoon waned, I turned back toward the hostel. My mind felt heavy with what I had seen, the beauty and the brutality, side by side. I felt grateful to have arrived safely but unsettled by how quickly comfort and horror could coexist here. That night, as I unpacked my gear, I began to think practically about what lay ahead: acquiring proper body armour, preparing for travel closer to the front lines, and arranging to meet the UK Aid team.

Before the real work could begin, I stopped at a small Mini Mart near the hostel. To my surprise, it was fully stocked, shelves lined with bread, fruit, and bottled water. Locals shopped calmly, chatting in low voices, their movements unhurried. Despite blackouts and air raids, life persisted. Supplies arrived daily from Poland and across the European Union, keeping the city nourished. Kyiv was battered, but far from broken.

That night, as the sirens wailed distantly and the lights flickered out, I sat by the window and looked across the quiet street. Somewhere out there, Max was preparing for the front, and I, in my small way, was preparing too. Tomorrow, I told myself, would be the day the real journey began.

CHAPTER 6

The UK Aid Team

Spending time with a UK Aid team was essential; as an organisation, we needed to know who they were and how they operated on the ground. With this in mind, I arranged to meet Andy in a remote location just outside the city, linking up with him during one of his aid runs. At first, it felt risky travelling alone to meet a man I had never seen before in an isolated place, but I knew I had to work around his schedule. He was, after all, helping us build vital connections in the region.

I booked a taxi, handed the driver a slip of paper with the address scrawled on it, and set off, unable to communicate without my electronic translator. When we arrived at the spot, there was no sign of Andy or the ambulance. The silence and emptiness made me feel suddenly isolated and uneasy. After a few tense minutes, I managed to get the driver to circle through a nearby housing estate in search of him. Then, as soon as I caught sight of him, I knew instantly; it had to be Andy: "a tall, broad figure, slightly balding, built like a skyscraper", just as he'd been described.

Andy, an experienced English ex-serviceman, had set up a front-line aid distribution facility in association with the Ukrainian Army. He had a compulsive personality disorder; this wasn't easy to adjust to at a practical level. From the moment you met him, it was challenging to speak without him talking over you or for him to initially consider any other point of view except his own. However, at the same time, he was a fantastic individual focused on getting aid to the people and front-line and Army battalions who needed it. I was keen to understand how the aid organisations worked as they were integrated within the military structure and how they functioned and Andy was the ideal guy to work with, I liked him. It was quickly evident that UK Aid was attached to a Battalion, the 112 Brigade Ukraine Defence Force. By association, you were also part of the Ukraine Military while working with UK Aid, this was something that I had not anticipated, but I had already begun to understand the need for it, you just couldn't move around the city or countryside without a letter of association from the Embassy or local battalion to present to the checkpoints in and out of the city.

From the beginning, working with UK Aid was full-on. Andy drove a modified ambulance, driving everywhere with the emergency lights on regardless of the job's priority through traffic in the emergency lane. We started to all work together and visited their local supply store on an army base near his home, sorting through the bundles of aid sent from the UK into the ambulance and then taking deliveries to hospitals and children's homes in Kyiv.

Andy and Olga his partner welcomed me into their home, offering a room and a mattress to rest on. Their house was a modern build, set back in a semi-rural stretch of land, surrounded by a tall concrete wall that gave a sense of both safety and privacy. The property sprawled across generous acreage, with more than enough room for vehicles, storage, and supplies. From an aid perspective, it was ideal, close to their distribution hub, secure, and practical in every way.

What stood out more than the property, though, was the atmosphere inside it. From the outset, our conversations were open, honest, and without pretence. I quickly came to value that straightforwardness. Andy, for his part, spoke often and with conviction. At first, I simply listened out of courtesy, but soon I began to realise how much I could learn from his perspective. His un-shakable support for Ukraine, his fierce dedication to Olga and their new family, and the blunt clarity with which he spoke gave me insights I might otherwise have missed.

Andy and Olga worked tirelessly from early morning, fuelled by countless cups of coffee, spending each day reaching out to as many people as they could and making as many deliveries as possible. The team's dedication and passion had become a beacon of hope for many locals, who had come to know UK Aid as a reliable and trusted partner. Andy, a well-spoken Englishman, was eager to learn as much Ukrainian as he could, while Olga, was originally from Siberia.

Fluent in Ukrainian, Russian, and several local dialects, she was the perfect counterpart to Andy, combining linguistic skill with a warm, understanding nature that made her a truly remarkable presence.

Mercedes ambulance painted in low reflected matt-black

The more time I spent with UK Aid, the more I liked them. They were dedicated, motivated and trustworthy. I began to enjoy the work in Kyiv, which was relentless, driving ambulances from one incident to another or just getting the tyres replaced to all terrain ready for the front. UK Aid had received an additional Mercedes ambulance to use. It was an all-wheel drive model equipped for the front lines painted in a matt black paint job to avoid attracting attention. It was startling to consider that driving an ambulance was perilous; the Russians would target ambulances or medics before other objectives.

The realisation that the Russians would rather target an ambulance than a legitimate military vehicle made me feel sick to my core, it was a stark reminder of their utter disregard for humanity.

Andy, Olga and their daughter Anna made a great team that had integrated totally into the Kyiv aid community with enhanced connections to the Ukrainian British community. Andy was a natural leader, a towering figure of a man over six foot, always on the move and seemed to be able to work with little sleep, ready to tackle the most challenging situations from the front. As long as you were willing to help and contribute, he was charismatic, easy to work with and supportive.

The more time I spent with Anna, their daughter, the more I admired her spirit. She was a 17-year-old teenager working through the challenges of adolescence during a full-scale war, trying to study, integrate with her community and maintain friendships. It was challenging at the best of times for her, a teenager, let alone for one who had limited time at college with patchy internet coverage and constant power outages. I initially had concerns about Anna as we left her at home most of the time while we were out. She must have found it daunting to be left alone in a remote location. With an increase in power cuts, she began spending most of the days isolated in the dark, feeling cut off, cold, and lonely. Her resilience and sense of humour impressed me. At home, I had experience working as a Chaplain in difficult situations with young people in drug rehab, homelessness, and violence. Under better circumstances, I had even worked at the other end of the spectrum, training youngsters in the

Australian Air Force Cadets (AAFC) as a Flying Officer, helping them survive and develop their skills. Yet, I still wondered how they would have faced the same challenges that Anna did. She was an exceptional young individual, capable, intelligent, and resilient.

It was also clear that while Andy and Olga focused on their aid work, they didn't forget about her welfare and were diligent parents. They set clear boundaries and maintained contact, ensuring that she knew she was cared for during her time on her own without them. At the same time, I was grateful for Anna's support and positive nature. It helped me stay focused and upbeat. If a 17-year-old could deal with what we were seeing, why shouldn't I be able to tackle it?

Day after day, we rose and proceeded with a clear sense of purpose. We dedicated most days to delivering aid to nominated hospitals, schools, or army medical units. The community's demand for support was immense. We had to prioritise the most pressing needs and evaluate each assistance request systematically.

One of the most rewarding deliveries we made was to a large children's hospital in the centre of Kyiv. After a recent shipment from the UK, we set about preparing baby milk, pushchairs, and vital medicines for children who were critically unwell. As we packed, I found myself watching Andy with interest. He didn't just load supplies at random; he paused, considered, and chose with care. His background as a veteran army medic was unmistakable. Each selection showed not only his knowledge of the medicines, their uses, interactions, and side effects, but also his compassion for the children who would receive them.

At first, I had underestimated the complexity of this process. Supplies often arrived from the UK unsorted, and the chaos of boxes piled high made it hard to know where to start. But with time and the introduction of large storage bins, we began to bring order to the chaos. Analgesics, antibiotics, and anti-fungals were sorted neatly; IV fluids, tourniquets, and advanced battle dressings were stacked where they could be found quickly. Standing in the storeroom, surrounded by shelves now carefully labelled, I felt a sense of quiet pride. We were bringing not just aid, but structure and reliability to a system that desperately needed both.

As time passed, I began to understand the importance of the UK Aid home and its location. The large rural property was enclosed by a solid concrete compound, with a formidable steel front gate that gave the site a sense of high security. From within, there was a clear view over the surrounding fields and scattered neighbours. It was an ideal base, safe, practical, and with ample room for future expansion. Several unused outbuildings were already being repurposed for storage and accommodation, a clear sign that Andy had a long-term vision and was steadily building capacity.

The team not only provided me with a room but also cooked an evening meal each night. In return, I did my best to contribute where I could, paying for lunch for everyone from time to time or covering the cost of fuel for the ambulances. It felt only right to give something back in exchange for their generosity.

Regular deliveries came in from the UK, anything from large forty-tonne lorries to smaller vans driven by volunteers; one was from DIY Ukraine, a Dover-based charity. Hector arrived with a young American who had just finished a ninety-day tour with a local Battalion; it was then that I began to also understand the lack of military training provided by the battalions. Corbin was a typical young American who spoke confidently but had become tired of how the Ukrainian Army had used him with little training and stuck him in the front line under artillery fire for weeks. I needed to understand how he had been recruited and spoke to him at length. The battalion he had served with was not attached to the Ukrainian Legion; the unit had picked him up on his way into Kyiv and recruited him directly into their ranks, giving him a couple of days of training with an AK76 rifle moving him to the front lines without any ballistic protection signing him into their ranks deploying him into the trenches. He was young and naive, keen to defend Ukraine against Russia, we had this in common.

He had completed a ninety-day visa-restricted tour and received only a basic discharge note at the end. When he showed me the paperwork written in Ukrainian, it looked more like a small concert flyer rather than an official discharge record. As I studied the document, it struck me that I too needed some form of additional paperwork to cover my own visit. I pressed Andy for a letter of association, choosing to travel under my British citizenship rather than my Australian one. A few days later, he handed me a simple letter of introduction linked to UK Aid. Though basic in form, it provided me with a crucial point of reference if ever stopped by a Ukrainian official at a checkpoint or border crossing while travelling in one of the ambulances.

The more I spoke to people like Hector and Corbin, the more I began to grasp the environment, people, and culture. The more individuals I encountered, the clearer it became why Ukrainians yearned for freedom from Russia and wished to maintain their independence.

Ukraine possesses an inherently rich, independent world-view with extensive expanses of productive farmland. While some older Ukrainians, who came of age under the USSR regime, might entertain the idea of communism's return, most of the younger generation were raised in a free and open society where they could chart their course and savour the taste of freedom. They had no desire to revert to the past.

President Volodymyr Zelensky is a prominent public figure who exudes charismatic national independence in an internationally unique manner through his nightly social media broadcasts. For the first time in history, Ukraine has a leader equipped to tackle Putin's regime on their terms. As I journeyed through Ukraine, connecting with its people and immersing myself in its culture, I began to comprehend how it had evolved since the Cold War and the collapse of the Berlin Wall. It is distinct and offers its citizens the freedom and prosperity Russia would struggle to provide. New buildings dot the landscape, and society is gradually embracing the Western way of life, along with all the complexities it entails. Although Russia ranks as one of the world's largest countries, it lacks the fertile farmlands of Ukraine, boasting over 40 million hectares of agricultural land, contributing 10% to the global wheat market, 15% to the corn market, and 13% to the barley market. Russia covets these resources.

Furthermore, I acknowledged that all aid workers and volunteers who journey to Ukraine to support their resistance against Russia constitute a unique breed. They are fiercely independent and comprehend that the conflict with Russia stands as one of the most pivotal struggles of our era. The free world and NATO cannot permit Russia to overrun Ukraine.

Hector was one of life's quietly confident characters, a hippy at heart living life to the full, happy with his place in the world to the point where he just wanted to enjoy life and travel Europe exploring all that it had to offer. He was the type of man who would give you his last sweet in the packet if you were fair with him. Hector was great to be around and went out of his way to support Ukraine, completing regular aid runs from the UK to Kyiv. From what I could figure, whilst he had a home in Dover, he spent most of his time working in different places worldwide, drifting from one event to another and exploring all Europe had to offer, sleeping in his van.

In my first weeks in the field, I discovered the brutality of the Russian invasion. The scars from their attack were visible everywhere I looked. Remaining near Bucha granted me an immediate understanding of the events of February 2022. During one of our visits to the UK Aid store to unload Hector's van, we received an escorted limited tour of the buildings near the store. One of these expansive, historically ornate buildings had functioned as a hospital or sanatorium for hundreds of people in the local community.

Upon entering the building, it became clear that we were witnessing the brutal reality of the Putin regime's scorched-earth policy. What had once been a modern hospital, fully equipped with the facilities needed to provide care and rehabilitation had been reduced to ruins. The structure itself was shattered, stripped of its ability to function.

In one of the first rooms we entered, a CT scanner lay dismantled. Its casing had been pried open, the control panels forced apart, and wires torn from their circuit boards, which bore the clear marks of hammer blows. The control room was gone entirely, electronic panels, monitors, and diagnostic systems ripped from the walls, leaving only scars where life-saving equipment had once been.

Witnessing such needless destruction was extremely frustrating. As we quietly navigated through the building, my sense of dismay intensified. On the second floor in the wards, all the patient's sinks and water taps had been shattered, leaving them inoperable. The building had reached a point of irreparable damage. The Russians had gone so far as to sever the water purification and oxygen supply networks to the wards. The water purification system had been deliberately damaged so that only a skeletal structure of pipes remained hanging on the walls.

We all felt the impact of the tour. There was a clear and deliberate purpose behind this destruction, carried out systematically, from the most essential items like sinks and side tables to the most advanced CT Scanner. The damage was total. Hector was quieter than usual, his typically upbeat attitude deflated. On the other hand, Corbin seemed to accept the destruction more readily. It wasn't until you understood that Corbin had been exposed to the Russians for months and had acclimatised to their relentless attitude to destroy anything Ukrainian, it was just another day in Ukraine for him.

I continued to work with Andy and Olga for days, trip after trip, delivery after delivery, until I also started to feel tired, run down and needed a change of pace.

It was at this point that another independent volunteer arrived, Gary, a retired policeman from the UK, Midlands, he had brought in another delivery. filled with bedding and medical products. He exuded hope and a strong desire to assist Andy and Ukraine. Gary had rented a van in the UK and enjoyed participating and delivering aid. Simultaneously, he had also organised the transportation of a refugee from the Donbas region to the UK, providing her shelter through a UK based refugee program. Despite the van's cost exceeding the value of its contents, Gary's focus lay in being part of the effort, rendering the expense irrelevant.

Cruise Missiles

After another four days at Andy's house, we were settling into a routine; by the time Gary arrived, the mood was busy and friendly. As he and I were considerably older than Andy, the UK Aid team named us Captain Mainwaring and Spike, characters from a historical series on TV called Dads Army.

Just after breakfast, Gary and I strolled up the garden from the house to several outbuildings on the grounds to inspect the solar panels on the roof of the buildings, checking if they were repairable. It was a refreshing walk up the garden in the crisp morning air. We relished our morning conversation, getting acquainted and discussing what we had in common together with our location, its beauty and suitability for its current purpose. It was a fantastic compound, perfectly positioned to access Kyiv whilst being isolated enough for privacy and security.

Suddenly, three objects shot into view from the south, racing across the sky screaming in to sight just metres above the rooftops. Before we could react, they had thundered over us in seconds, one after another, almost knocking two builders off the roof of a house just to the south of our location.

Downed Russian, Cruise missile just outside Kyiv

Seeing three cruise missiles slicing through the sky overhead was staggering. This chilling spectacle left us utterly flabbergasted as they raced northward above us, their deadly trajectory set on Kyiv. We just stood in the middle of the compound looking at each other, not knowing what to say and laughed. I don't know why we laughed; the situation seemed unreal. There were now air raids on Kyiv daily, and these three projectiles were just part of the day's round of attacks.

This type of spectacle became part of everyday life; the air raid alarm would go off on your phone each day, providing details of the incoming projectiles. The information on your phone gave you details of the firing location, direction of travel and estimated time to impact. Ukraine's sophisticated air defence system was now intercepting many of the missiles. In some cases, we could even hear ground fire targeting the projectiles in the distance. Hiding in the air raid shelter during every raid was impractical; there were two or three raids daily, mainly in the morning. If we had hidden, we would never have accomplished anything.

We continued with the day and discussed the event later in the day. I had never considered living to see the day I would have seen so many cruise missiles and air raid attacks. I struggled through it briefly to the extent that I ignored the risk and ongoing attacks, but I could see the effect on the population and people around me. Understandably, it affected some people more than others, particularly the younger and sensitive community members. As the air raid warnings sounded, you could instantly see the fear in their eyes and the panic among them. It was physical and changed the mood in the room or area where you were immediately. I felt lucky that it didn't impact me at the time, but understood later that it would come back to haunt me.

For the local people, this place was home, a home defined by relentless attacks and the unending shadow of air raids. I was only a visitor, and though the violence left its mark on me, I found solace in knowing I could return to safety, an option they did not have.

The Zoo

Over the next couple of weeks, the team's schedule became busier, inundated with more help requests than they could manage. Gary, still present, volunteered to conduct an afternoon aid run in his van. Andy appreciated the gesture and was eager to take an additional trip to a distant animal farm sheltering tropical and rare species in Ukraine. I found this somewhat amusing, considering that most of the animals, parrots, wombats, and wild cats were commonly found in the wild in Australia.

The farm had lost half of its species when the Russians invaded through the Belarus border and were cut down together with their owners. The survivors were slowly starving and lacked the specialist needs for their breed. They desperately needed food, bedding and medicine. Gary and I were given the zoo's location and loaded the van quickly. Andy and Olga had already left on a separate hospital run, and Anna decided to come with us to navigate and translate. We did need a translator and were grateful for the thought. I hated going places without a translator; even though the locals were grateful for the help, they were war-weary and impatient. Anna, had effectively become our 17 year old "fixer".

We set off after lunch after a quick bite of food and headed north with Anna guiding us with Google Maps. It was cold and drizzling with a low haze hanging in the air. As we drove up the E373 road towards Belarus, we played a game counting the number of destroyed and burnt-out pieces of armour littering the road. There was a jolly atmosphere to the trip, and we began to take our focus off the reality of the journey. It wasn't long before I became aware that we had been on the road for ninety minutes, and the trip should have only taken us sixty. As I travelled everywhere with my GPS navigator, I powered it on and got an "actual" North-South location; "shit", I went cold and felt sick; we were only twelve kilometres from the Belarus border. I told Gary to stop the van and turn around. The sudden change in my demeanour made no impact on him, as he seemed as happy as ever, initially ignoring my request, even to the point of questioning the accuracy of the GPS, preferring to trust in Google Maps. I suddenly felt scared and unsettled, the kind of discomfort that forces you to become assertive. Gary needed to stop before we risked driving straight into the sights of Ukrainian border guards, or worse, a Russian checkpoint. For the first time since we'd met, he could see my mood had shifted dramatically. I was no longer calm but visibly agitated, even bordering on aggressive, something that must have seemed completely out of character to him. With a questioning look, he pulled over so we could reassess our position.

As we pored over the map, Gary's own agitation surfaced. The daylight was fading fast, and it was obvious we were running late; by now we should already have arrived. Anna, amazingly composed for a 17-year-old, phoned her mother to confirm the address. The truth landed heavily; we were nowhere near where we should have been and were, in fact, edging dangerously close to the Belarusian border.

Frustration welled up inside me. I was angry with myself for not relying more on basic field-craft. Too many people had already lost their way by trusting Google Maps alone, unable to tell north from south. With Anna's help, I recalculated the route, keeping the GPS on and fixing our location.

I spoke quietly to Gary, trying to ease the tension. Once he realised where we were, embarrassment and anger flickered in his eyes. He had always seen himself as an experienced professional, and the misjudgement cut deeply. The only thing left was to make light of our mistake, take the lesson, and push forward, quickly and without fuss.

Although Gary had been a little too "happy-go-lucky" and let his attention slip, he was a decent man whose company I genuinely enjoyed. Wanting to recover the situation, I reassured him that it could have happened to anyone, and at least we had realised our mistake before any real harm was done. With that, we both took a deep breath, steadied ourselves, and pressed on. We turned back and continued towards the zoo for another half an hour arriving as darkness fell.

The locals were pleased to see us and quickly jumped into action to help unload the van. By then, I had switched into a more professional mode, intent on getting the trip back on track and conscious of my responsibility for everyone's safety, especially Anna's. I gently took charge, eager to complete our task and return home.

Anna, though, was delighted to be at the zoo and wanted to see the animals. I had to pause, take a deep breath, and weigh the risks. I trusted the locals, but the fact that Gary and I couldn't speak the language left me uneasy. This time, Gary spoke to me in a gentler, more considerate tone, urging me to relax and join the tour. It didn't take long to realise that going along was not only polite but the right thing to do.

Walking through the partially deserted buildings, the zoo owners explained that their main concern was keeping as many of the surviving animals alive as possible so they could rebuild after the war. As we examined the enclosures, it became clear that most of the remaining mammals were solitary, unable to breed, and reliant on specialist diets. The most exotic were the wild cats, such as the lynx, that required large quantities of high-protein meals, an expensive challenge in this part of Ukraine.

Most of the tropical species were birds, particularly parrots and parakeets, including some native Australian varieties. Among them were bright pink and grey parakeets that instantly reminded me of home, where flocks of them would sweep across the skies during spring and autumn. The tour offered an unexpected sense of comfort and light relief from the harshness of the day, and I noticed how

Gary and Anna seemed to rediscover a childlike, playful spirit in that moment.

We could only hope that the supplies we had brought would help the zoo survive the coming winter and secure its future.

As we wandered through the outbuildings and enclosures, we snapped photos of the animals to be shared on social media by the team. At the same time, we spoke with the zoo owners through Anna. The more they revealed, the heavier my heart became. When the Russians invaded, they had not only targeted the exotic animals but had also killed many of the owners' family members, including women and children. Their story was a harrowing narrative to absorb.

Adjacent to the zoo lay a modest forest where the family members and animals were laid to rest. The graves were recent, with signs of freshly disturbed earth. Adding to the grimness, we were shown cleverly concealed shelters for the children amidst the graves. This unsettling reality emerged from the need to protect the children in the event of another Russian invasion. The concept of concealing them among the graves was a clever strategy to ensure they remained unnoticed. These were compact subterranean compartments stocked with a few days' worth of supplies and camouflaged air filters. As we walked away, I couldn't shake the image of children huddling together underground for days on end in cramped, suffocating conditions, "entombed". What haunted me most was not just the hardship, but the fact that families had felt compelled to build such hiding places. The Russians were brutal and unforgiving, capable of openly murdering, raping, or crippling people for sport and they were only forty kilometres away. To imagine trying to protect one's family under those conditions was nothing short of terrifying.

Walking around the zoo had been enjoyable, but the real purpose of the visit was to help ease suffering and share in the lives of those we encountered. I began to understand why so many felt the need to tell us their stories. The people we met wanted the world to know how they had been treated by Russia and that their deepest desire was independence. Once they learned we were from the UK, the Commonwealth, or Australia, they made certain we heard as much as we were willing to listen to. Often, they would go further, urging us to ensure their truths were shared with the world, asking us to act as witnesses to crimes against humanity.

By the time we had finished and completed the tour, we all left with a renewed sense of purpose and another story to tell.

CHAPTER 7

Ukraine Patriots

During one of our frequent visits to the UK Aid stores concealed on the nearby Army base, the Canadians arrived to help organise our medical supplies and replenish their team. At first, I was told little about them, only that they were naturalised Ukrainian-Canadians living in Kyiv.

Andy had instructed me to limit my conversations with them. Initially, I assumed this was for security reasons, but soon realised it was UK Aid's way of managing my accessibility to Ukraine Patriot.

That morning we worked through unopened boxes and crates, sorting medications into dispensing containers. Their team included a qualified pharmacist who could quickly identify everything, including advanced medicines, making the task far smoother. They were easy to collaborate with, and their fluent English made communication refreshing. They were naturalised Ukrainians from Canada who resided in Kyiv and had chosen to stay and resist the Russians during the full scale invasion in February 2022.

Their leader, Lana, had founded an aid agency called Ukraine Patriots. Her aim was to develop connections with UK Aid and build networks, seeking opportunities for collaboration in future capacity-building endeavours. In essence her goal was similar to mine, capacity building in the aid community.

During a quieter moment out of sight, I talked to her directly; she was a feisty young woman who spoke frankly without wasting time. She asked me who I was and why I was in Ukraine. I quickly explained that I was from Australia in Kyiv in a civilian capacity, looking at areas where Australia could support Ukraine at a grassroots level in association with the Embassy and Chaplaincy Australia. Lana struck me as genuine. I discreetly gave her my number, and we arranged to meet again.

A few days later, we sat in a quiet hostel tea room in Kyiv. Once she was satisfied with my credentials, Lana opened up. She told me of her career as a ballerina with the Virsky National Honoured Academic Ensemble of Ukraine, an extraordinary achievement, and one that explained her charisma and determination. I began to understand why Ukraine Patriots had grown so quickly under her leadership.

Listening to her describe their work, I realised this was exactly the kind of partner Australia had been searching for: a Ukrainian-led, trustworthy group capable of getting aid into the hardest-hit regions without losing consignments to rogue elements. She spoke of "spicy runs", aid deliveries into front-line areas under artillery threat, navigating destroyed infrastructure and constant danger. Their work was defined by urgency, compassion, and local knowledge.

Convinced of the potential, I drafted a report for the AKU (Australia), recommending Ukraine Patriots as a key partner. Introducing Lana to the Australian community and the AFUO (Australian Federation of Ukraine Organisations), felt like the right next step, a chance to build something lasting that could truly help Ukraine's most vulnerable communities.

The Hostel Team

After a few weeks with the UK Aid Team, I was ready for a change. I knew that there was a lot more work to complete. Andy, Olga and Anna understood my need to move on and connect with other organisations, so I thanked them for their hospitality and insight and moved back to the hostel in Kyiv just as the air raids across the city increased and the power cuts began regularly. Russia had recently targeted Ukraine's power infrastructure, overwhelming its capacity to meet demand. Consequently, Kyiv's mayor enforced intermittent power outages throughout the city. While most of these outages followed a schedule, some occurred unexpectedly. We had to swiftly adjust to this fresh challenge by keeping candles and head torches ready. It was possible to be en route to the bathroom at night only to have the power abruptly shut off, plunging you into complete darkness with no sense of your surroundings. Total darkness to the point where you couldn't see your hand in front of your face. Carrying a head torch at all times became essential.

The Dream Hostel, candle light during a power cut, Air Raid

The atmosphere at the hostel had changed; everyone was on edge, particularly the younger staff members. Most people had the air raid app on their phones, which went off in sync, creating a sense of panic in the building. I refused to give in to the fear and tried to sit confidently in the lounge or main hall, chatting with other residents under the glow of candlelight.

As I had more time, I started getting to know more of the residents. The mix was eclectic: seasoned humanitarian workers moving quietly and efficiently between briefings; hardened journalists chasing fragments of truth through shellfire and rumour; and, inevitably, a scattering of individuals who had drifted in as self-styled war tourists. The former understood the gravity of where they were. The latter treated it, at times, as theatre.

Among the most compelling were the European and American

volunteers serving with the International Legion of Ukraine, formally known as the International Legion of Territorial Defence of Ukraine. Many were former soldiers from across the world, measured, disciplined, and surprisingly thoughtful. Away from the front they were reflective rather than boastful, analytical rather than dramatic. They spoke in practical terms about logistics, terrain, artillery patterns, and morale. Their understanding of the war was not abstract; it had been earned.

The press, by contrast, were treated cautiously. Operational security was a constant undercurrent, and trust had to be built slowly. Yet in private conversation, many journalists proved to be deeply informed and perceptive, often carrying a broader geopolitical perspective than those of us absorbed in the immediacy of convoy routes and supply manifests.

The least respected group, however, were the war tourists and the young men who loudly claimed affiliation with intelligence services, some declaring themselves connected to the CIA, others to MI6. It was baffling. In a war zone, discretion is survival. No legitimate intelligence officer would ever announce such a role; to do so would be reckless, not only for oneself but for everyone nearby. Declaring such an identity in that environment was not bravado, it was Naïveté bordering on dangerous fantasy. Having only recently met Vasily in Lviv, where the atmosphere carried its own quiet tension beneath the café chatter and railway announcements, it had already become clear how easy it was to identify the tourists. Their posture gave them away, the preformative intensity, the conspicuous clothing, the over-eagerness to be seen. They were usually, and politely, asked to move on. Ukraine was not a stage. For those who were there to serve, to report truthfully, or to fight under lawful command, the war was neither spectacle nor identity, it was becoming a responsibility.

At first, I struggled to believe that there could be such a thing as war tourists, men from around the world who came to Ukraine merely to experience the atmosphere of conflict, without offering any help or contributing to the resolution of its problems. Yet, as I soon discovered, this was only the beginning. I was about to encounter some of the more unsavoury characters who had embedded themselves within the community.

Occasionally, a couple of charismatic, physically fit soldiers in uniform would visit the bar in the evenings searching for the war tourists. Whilst I was disgusted with the tone of their conversations, I felt compelled to listen to their proposition finding myself overwhelmed by the content of their tour business.

The two visitors were special service veterans looking for work. While they did offer genuine close protection and had valid security credentials, their costs were extortionate and well out of our price range for visits to the front line. They too also offered "slot tours".

At first I found it difficult to comprehend what a "slot tour" was. The explanation was simple, the Americans were licensed contractors, employed to provide close protection to dignitaries. To make more money on the side they offered their services directly to anyone who wanted to go to the front lines with a high calibre weapon to kill a Russian. Whilst they were confident with their abilities and reassured anyone who showed an interest that they would be safe, providing them with all of the protective equipment possible, it was an expensive way to spend five hundred dollars a day to kill another human being. Quiet frankly, the idea was "repugnant" but we all needed to understand that we were at the mercy of the conditions of total war. Rather than display my disgust, I tried to ask them for more details of the opportunity trying not to come across as "too" interested.

However, I was sure that my body language would have given me away as I retorted in my seat as they spoke about some of their ideas.

In the meantime, one of the most interesting people I met was an independent Scottish reporter, Jen Stout. We spent a couple of evenings together in a group discussing the war during blackouts and air raids. Jen had grown up fascinated by Russian culture, spending a considerable amount of time in the region, writing articles about the suffering of the people and her associates in Kharkiv, who had become her friends over time. She was worried about their safety as they were bombarded every day by artillery and deliberately kept awake at night.

Jen had a unique perspective on the situation in Ukraine because she knew the people and understood their culture. She even spoke the language and various dialects. Like me, she believed that Ukraine needed more long-range tactical weapons to counterattack the Russians and level the playing field.

While she had been fascinated by Russian culture, Jen had begun to understand that the Russian political far right saw the Ukrainians as almost subhuman Nazis who had no right to exist and should be either re-educated or executed. There was no doubt in her mind that Putin's regime held the same feelings towards Moldova and Poland, with an eye on their borders, harbouring the same hostile sentiment towards them. All the indications pointed in that direction: Putin had ambitions beyond Ukraine's borders. This is why Ukrainians declared that they were fighting for peace and freedom in Europe. They are the cutting edge of the war and fighting it on our behalf.

To help us through the dark, we would buy drinks at the bar, light candles, sing along, or chat about each other's interests and backgrounds. At the top of the agenda was how Ukraine would win the war. "Slava Ukraine" was a popular subject and always made people smile. However, the harsh reality of the situation and overwhelming power of Russia was always there in the back of your mind.

As relationships flourished within the hostel, we established additional connections and organised more aid runs. I started venturing out with more individuals I had developed trust in, even joining press members and military affiliates on visits. In essence, the hostel was being used as a hub by aid agencies and battalion commanders to recruit soldiers, drone operators and medical staff. There were resident representatives from many European countries and America. The more you got to know them, the more you began to understand the situation in Ukraine and the political landscape, particularly the desire to move from the post Soviet era to a European country. Most people I spoke to genuinely wanted to be independent of Moscow and were scared of Russia and the old USSR regime of the 1980's.

Bucha Massacre

Staying away from Bucha the town was difficult as we drove through it most days, sometimes six or seven times a day. It didn't take long before I began to connect with the locals and hear awful stories of the atrocities they had experienced.

During the full-scale invasion of Ukraine in February, one of Russia's initial moves was a push toward the Ukrainian capital, Kyiv, with a massive column of military vehicles advancing from Belarus. The Russians came close to entering the outskirts of Kyiv before they were overwhelmed and halted by the heroic efforts of the Ukrainian Defence Force.

Among the Russian units that spearheaded the Battle of Bucha was the 234th Russian Air Assault Regiment, one of the formations named as a perpetrator of the massacre along Yablunska Street. The unit had been led by Lieutenant Colonel Artyom Gorodilov. Evidence shows that the killings were part of a deliberate and systematic effort to secure a route to Kyiv, executed with ruthless precision. Soldiers interrogated and executed unarmed men of fighting age and killed civilians who unknowingly crossed their path. Amongst the dead included children fleeing with their families, locals out searching for groceries, or people simply trying to return home on their bicycles.

What I was about to witness made it clear that these killings were no accident. They were part of a brutal and methodical campaign to eliminate any perceived obstacle or witness.

Soldiers had interrogated and executed unarmed civilians, including men of military age, but also killed anyone unfortunate enough to cross their path. These victims included children, families, the elderly, and those simply trying to survive amidst chaos.

On one particular day, I had gone on a rare trip to the town centre to buy groceries when an elderly couple approached us. They insisted we accompany them to meet a local official near City Hall. Initially, I was reluctant. We received many such requests, and it was difficult to honour them all.

Max, our interpreter, spoke quietly with the couple for a few minutes before returning to us. He urged us to go with them, explaining that they wanted us to see historical evidence of the massacre. The couple had been adamant. They needed as many overseas, independent witnesses as possible to verify the atrocities and help spread the truth across the world.

It became clear that this couple had lost their entire family in the invasion. Their grief had transformed into a mission to ensure the world knew what had happened in Bucha. We were the first group they had encountered that included an Australian, which made us particularly valuable as independent witnesses to war crimes.

We loaded the couple into our van and drove to a building that resembled the local town hall. There, they introduced us to a city official who was stationed behind the main counter.

He led us inside and down a set of stairs into a large, tiled chamber that had been transformed into a makeshift mortuary and records archive. This room contained remains and documentation, evidence of the massacre. Some of the bodies and records came from a basement beneath a nearby campground, allegedly used as a torture site during the February 2022 invasion.

There were dozens of statements and photographs detailing the unlawful killings of at least seventy-three civilians in one room alone. The photos were horrifying, bodies of civilians, hands tied behind their backs, shot at point-blank range.

The remains in the mortuary confirmed the atrocity. Many of the victims were young women and children, their hands tied, their bodies burned in an attempt to destroy evidence. The trauma in that room was indescribable. I felt nauseated, struggling to focus on the conversation. The stench of death, charred remains, and suffering was overpowering. Every instinct I had was telling me to run.

"Run." "Run, now."

But I stayed. I had made a commitment to witness and document the atrocity, and I forced myself to find the strength to continue. Just as I began taking in the scene, the elderly couple began sharing the most personal, painful parts of their story. They held out photographs of their murdered family, speaking softly about each person, their jobs, their school achievements, their lives.

As their testimony became more intimate, its emotional weight settled on all of us like a physical pressure. I could feel the group faltering under the burden. Even Max, who was part Ukrainian and usually composed, had turned pale. I had never seen him like that.

Some victims had been burned beyond recognition, others mutilated. These were the unclaimed and the unidentifiable. DNA testing was the only way to eventually return their names to their families, and the wait for that confirmation was agonising.

We walked through the chamber. I tried not to look, but it was impossible not to see the charred and disfigured bodies of women and children. These had been innocent lives, executed, tortured, discarded. The local official guiding us was visibly affected, his composure cracking under the weight of what he had to relive every day.

The rooms were cold and damp, filled with a sense of deep, human catastrophe. As I listened to survivor after survivor share their stories, each one more horrifying than the last, I couldn't hold back tears. Looking around, I saw that I wasn't alone. Everyone in our group was struggling.

The photographs we were shown were devastating, civilians lined up, their hands tied behind their backs, executed at close range. The trauma was palpable. For the people of Bucha, it was not a memory; it was their everyday reality.

Eventually, I reached for my notebook. I knew I needed to take notes, but my hands shook, and I didn't know what to write. We were overwhelmed.

Still, the locals needed us to understand. I began asking questions through our translator and jotted down what I could, determined to bear witness and later share an honest account of what we had seen and felt.

The visit lasted for hours. The medical assistant working in the morgue was passionate, and walking away before he had finished showing us everything would have felt disrespectful.

So I stayed. I listened. I tried to comprehend the immense burden he carried. But his suffering was not isolated, every resident in the town bore similar emotional scars, all shaped by the same horrific event.

Max and I assured the family that we would share their story and try to engage Australian officials. We would make sure the world knew what had happened in Bucha.

As we left, the weight of the visit hit me physically. I vomited into a towel in the back of the ambulance. When I looked at Max and the others, I saw the same pale, stricken faces. Everyone who had seen the basement carried the trauma with them.

We drove back in silence, each of us lost in our thoughts, unable to speak about what we had just witnessed.

In an attempt to break the tension, we stopped along the main road in Bucha to take photographs of the burnt-out Russian armour left behind from the assault. For a brief moment, we smiled, and then guilt washed over us. We were smiling in front of symbols of death, destruction, and pain.

Later that day, the full impact of what we had witnessed began to settle in. The experience had traumatised all of us.

Decimated armour littered the E271, from Bucha to Belarus.

The horrors I saw were incomprehensible. I could not understand how innocent civilians, especially women and children had been subjected to such cruelty and violence. I knew instinctively that I would need to compartmentalise this trauma, to lock it away in my mind until I could begin to process it.

But I would never forget.

That day, I vowed to stand as a witness to the crimes against humanity I had seen. I pledged to stand with the people of Ukraine and to share their truth with the world. What happened in Bucha was not propaganda or exaggeration, it was real. It was a massacre. A deliberate atrocity, carried out by the Russian military machine and sanctioned by the leadership in Moscow. The fallout from Bucha will remain with all of us forever, a stain on our collective conscience and a reminder of what happens when brutality is allowed to march unchecked.

CHAPTER 8

Kherson

On 11 November 2022, Ukraine reclaimed control of the city of Kherson and adjacent areas in Kherson Oblast, along with portions of Mykolaiv Oblast on the right bank of the Dnieper River. The Russian Armed Forces executed a withdrawal to the left bank between November 9 and 11, November. These developments transpired due to a southern counteroffensive from September to October 2022.

Earlier, Russian General Sergey Surovikin officially declared the withdrawal of troops from Kherson and the western bank of the Dnieper River. He justified this decision by pointing out the insufficient supply capacity to Kherson and its neighbouring settlements across the river due to intense targeting by the Ukrainian forces. Additionally, he underscored the risk posed to civilians due to ongoing shelling.

I had integrated well with the aid community at the hostel, as well as with the young conscripts and soldiers who stayed there. While most of our intelligence came through official channels like the embassy and government agencies, local sources proved equally invaluable.

As our understanding of the situation in Kherson deepened, Max was tasked by his battalion commander to organise a civilian aid run to address the community's immediate needs following the city's liberation.

I had grown fond of Max, having bonded with him from the very beginning, our first encounter over breakfast, when he had casually shared his meal. That moment seemed to capture his character perfectly: generous, open, and quick to make friends. I trusted him and enjoyed his company; he was both reliable and good-humoured. Though I had been reluctant to venture so close to the front line and the Russian positions, Max managed to convince me.

After I decided to go to Kherson, I understood that it could mark my final aid run and my last chance to establish relationships or discover additional contacts for the team during my first deployment. We had already spent a week planning while we completed other visits, deliveries and driving ambulances. Although I was unwell and hesitant about participating in the run, my local teammates persuaded me to join them, as they desperately needed a fourth person willing to take a turn driving.

The ruins of Kherson during November 2022

It was common knowledge that this mission was locally dubbed a "spicy run," indicating that it would take us into a combat zone. While these missions carried the highest risks, they also yielded the most significant rewards because the civilians in these areas teetered on the brink of desperation and were in dire need of aid, food, bedding, and medical assistance. We had a medic with us and one van packed with medicine and critical IV fluids; the second vehicle was primarily loaded with food and baby diapers.

Gathering all the necessary paperwork for these runs consumed the most time, as it involved navigating the bureaucratic requirements to travel across the country and through various security zones. Access became progressively more challenging as we approached the front lines, which was entirely understandable. Fortunately, Max had additional ties to the legion, which expedited the paperwork process. I had to rely on the team's assurances regarding the paperwork, as I couldn't read most of it, as it was written in Ukrainian.

The third requirement was securing a translator—a local who not only understood the language but also the people and terrain better than Max ever could. Fortune was on our side. Among the refugees staying at the hostel was a woman from Kherson, displaced by the fighting but determined to return home. Her name was Nastasiya.

From the beginning, it was clear she was more than just a translator. She carried with her a fierce commitment to her community and a longing to be reunited with her family. Though it was difficult, we persuaded her to leave most of her relatives behind in Kyiv for the time being, assuring her that once she had re-established herself in Kherson, they could safely follow.

Her value quickly became apparent. Nastasiya was deeply embedded in the fabric of her region, she knew the district's geography, its rhythms, its people. More importantly, she was familiar with local dignitaries and even the battalion commander, connections that would prove vital in ensuring safe passage and building trust. She was not just bridging our words; she was bridging cultures, smoothing over suspicion, and opening doors that might otherwise have remained closed.

In her, we had found not only a translator but a guide, a "fixer", and an ally, someone whose personal stake in the fate of Kherson aligned with our mission.

We embarked on our journey bright and early, just after four a.m. We swiftly loaded the large vans from a local warehouse and headed south. Max and Nastasiya, became our local, "fixers" and translators, with Max accompanying us in the first van, while I took the wheel with Ben another member of the hostel team and Nastasiya in the second. All of us carried flasks of hot tea and biscuits. Most of the biscuits and trinkets in the cabs were intended as handouts for the checkpoint guards, a mutually understood practice that facilitated our passage through the roadblock.

As we left the city and entered the suburbs, the roads were lined with checkpoints. Guards stood before bunkers or sheds, many constructed using handmade structures and barriers crafted from pallet wood. We had to navigate remnants of concrete dragons' teeth, mostly left over from the initial invasion.

Most of our team had worked in the region for months, and their experience proved invaluable. We had adorned the vans with medical crosses akin to ambulances, which never failed to elicit smiles from the checkpoint guards as we approached but did increase the chance of us being attacked by the Russians. The atmosphere among us was filled with confidence, which lifted my spirits and propelled me forward.

Offering biscuits and glancing at the paperwork with "Slava Ukraine" swiftly got us through most checkpoints.

Once we left the city and reached the main roads, our progress slowed as the roads had deteriorated, bearing the scars of past impacts. Nonetheless, we settled into a smooth rhythm, occasionally stopping at several service stations for breaks and supplies. Another rule for these runs was to ensure we had ample fuel in case supplies were limited in Kherson and enough warm clothing and bedding to sleep comfortably in the vans.

As we approached Kherson, the inspections and conversations at the checkpoints became more frequent and thorough. It was well into the afternoon before we finally reached our agreed-upon destination and connected with our contact on the Kherson outskirts.

Ukrainian soldiers standing on a checkpoint

We had been actively collaborating with various agencies, engaging in daily visits to numerous civilian aid and military locations. Distinguishing between the two was challenging, given that most civilian aid organisations maintained some form of connection with military battalions. This connection was essential to facilitate the movement required for aid distribution. These relationships were mutually beneficial, aid agencies relied on the support of battalion commanders to navigate, while the battalions gained the advantage of receiving priority access to the delivered aid.

Handing out aid in a conflict zone was an incredibly challenging endeavour due to the chaotic and potentially volatile conditions. Most of the civilians who had stayed in Kherson were desperate, starving and had been treated with contempt by the Russians brutally. To maintain control and ensure the efficient distribution of aid, we sought assistance from the local army, working closely with their commanding officer and our local contact to establish order.

By now, the AKU and Embassy network had significantly grown, enabling us to gain local intelligence actively through the multiple chat channels on the Signal and Telegram apps. While we occasionally used WhatsApp for chat, we were aware of its vulnerability to spoofing or hacking by other agencies.

Most of the local intelligence was fresh and up-to-date, continuously refreshing in real-time alongside official news. We had to consider a lot of information. One essential reminder was to ensure that we switched off our phone's GPS location function to avoid the enemy picking up or pinging our location. It was surprising how many aid workers forgot to do this before heading out on the run or only turning their phones on when necessary.

In this environment, the safety and security of both us the aid workers and the recipients was paramount. The local army played a crucial role in providing security and crowd management.

Together with the local commander, we devised a plan to organise the public into orderly lines, reducing the risk of stampedes or confrontations as aid arrived. One van carrying medical supplies was directed to the local aid station, while the other, loaded with food and general provisions, drove to a central distribution point. Soldiers established a secure perimeter around the area, allowing us to work without undue risk. They also assisted in managing the flow of people, checking identification, and ensuring that aid was distributed fairly and equally.

Our intention was to prioritise the most vulnerable within the crowd and provide targeted support. However, it soon became clear that we had overlooked a critical element; many of the sickest and most frail were unable to come forward at all. The elderly and the infirm, crippled by illness or injury, remained hidden in the shadows of damp, underground bunkers, waiting for someone to find them.

Cooperation between humanitarian organisations and the local military was essential in this type of conflict zone, as they possessed crucial knowledge of the local terrain, dynamics, and potential security threats. Without them chaos could ignite instantly. The collaboration allowed us to reach those in need while mitigating potential risks and maintaining a degree of control in an otherwise overwhelming environment.

As we pulled up ready to hand out the aid, more and more people arrived, flooding out of destroyed buildings and basements. As the word spread more people arrived to the point where we knew that we would run out of aid quickly wishing that we had bought more. Within an hour the aid had been handed out and we moved on to a safer location to talk more to the local commander and return with a list of their priority items.

It is hard to believe that people still lived in the destroyed buildings and started appearing from the basements after we arrived. As they heard the vans pull up, spurred on by hunger and the hope of a warm meal, they stumbled out of the ruins covered in grey concrete dust and dressed in filthy clothes, stumbling over the debris trying to get to us first to see what we had to offer. They were tired and grateful for anything we could provide; the thought of a warm cup of soup and some bread made many of them cry and stroke our arms in the grateful gesture. These people of Kherson had nowhere else to go, from mums with infants to the elderly who could hardly walk, let alone navigate rubble-covered pavements to get to us. The landscape was surreal, like something from an apocalyptic movie, hard to process and understand. I could see the stress in their eyes and sympathise with their need for food, water and clean clothes.

There was no running water or sanitation; many people had open sores and smelt as if they hadn't washed for months. We could only stay briefly in the open and did what we could before moving to the following drop-off location.

Though the task remained difficult, the assistance of the local army and the coordination with the local commander were vital in ensuring that aid reached the intended recipients and contributed to stability in the conflict-affected region.

One of our other objectives was to evaluate the primary needs of Ukraine from an Australian, American and UK aid point of view. The AKU aimed to establish connections between aid agencies, army and drone manufacturers in Australia, facilitated through Damo and the Ukrainian Embassy in Australia.

During our aid run to Kherson, the aid team had unexpectedly informed the local battalion commander of my enthusiasm for drones. I had been working with the Australian Defence Organisation for the past seven years part time as a volunteer and had a genuine interest in drones, however, I really didn't want to get involved in a combat situation or visit the front-lines. I had no desire

to be involved with the military directly, I had gone to Ukraine as a civilian aid worker.

Hobby Drones in Warfare

Before I could negotiate my way around it, I had been introduced to Nikolai, the Battalion's drone instructor. Over the next two days, we collaborated on modifying Mavic 3 DJI drones, enhancing their capabilities by updating their motherboards and eliminating geofencing restrictions through BIOS (software) replacement. This alteration empowered them to fly without limitations and gave us a good idea of the components that they required for further development. This turned a commercially available off-the-shelf drone from a hobby item into a weapon of war, fully customisable and able to run surveillance or deliver fatal payloads.

I understood that the personnel at the local drone training centre possessed remarkably advanced equipment and capabilities. Their technicians were even constructing custom drones tailored explicitly for specific missions. These ranged from smaller swarm drones capable of agile, high-speed manoeuvres to larger multi-rotor units, some measuring up to three metres in size and carrying heavier payloads.

Most of these drones also featured remarkable capabilities, possessing spectral vision suitable for nocturnal operations and the capacity to carry a range of diverse payloads. They mainly employed smaller drones for surveillance and targeting objectives, while they modified medium-sized publicly available DJI drones with weapons for use in the field.

As I spent more time with Nikolai, it became clear that the community viewed drones as an essential part of the aid required for Ukraine's defence. To them, there was no distinction between the supply of drones and the delivery of warm blankets or food. I soon realised that I had been deliberately side-tracked by the battalion commander and paired with Nikolai to emphasise their urgent need for more drone equipment, as well as counter-drone defences against the Russians. The only reason I had been chosen for this task was my background in electronics and aviation, which allowed me to understand the technology. With a few days still remaining before our departure, I had little choice but to follow their lead, shadowing Nikolai closely and taking careful notes on the emerging technologies and processes being developed.

I was pleasantly surprised by how easily the drones could be modified. Drawing on my background in electronics and computers, we guided the team through straightforward instructions sourced online. We replaced the motherboards and upgraded the antenna modules to higher-gain versions. These adjustments dramatically expanded the drones' capabilities, allowing flights of up to eight kilometres and altitudes of two kilometres or more. The pace of change was remarkable, new ideas and creations seemed to emerge every single day, a constant evolution that underscored just how fluid and adaptive the situation had become.

Typical Platoon mobile drone unit

Moreover, we successfully tweaked the drone firmware, allowing us to assign custom names to each drone in flight. This strategic alteration helped us conceal the drones from Russian scanners, complicating their ability to discern the ownership of these drones, whether they belonged to the Russians or the Ukrainians.

It only took us three or four hours to update a Mavic drone ready to carry ordnance. In most cases, with modified propellers with a greater angle of attack and lift, a Mavic would easily hold a grenade that weighed about 600 grams. A modified actuator underneath was used instead of an existing drone command to release the grenade when ready. By now, I had been as close to the Russians as I had ever been, only a thousand metres away across from their front lines. I was unhappy; I had not gone to Kherson to work with the military but the Embassy was interested in developing commercial links to help connect Australian developers to Ukrainian units.

By now, Nikolai had dragged me into a forward position to show me how the drones worked in the field. The noise of war was constant, a cacophony of harsh sounds jarring from one to another, deafening at times overlaid with each other, unable to decipher the sound of incoming to outgoing artillery and explosions with small arms fire constantly overlaid. Half the time, you needed to decide whether to take cover or move. However, you knew when you were being directly targeted as the sound of "whizzing" bullets passing your head caught your attention as you instinctively hugged the ground and looked for cover.

We now moved from trench to trench in two-person teams with a drone, a couple of charged batteries and a handful of grenades in a separate bag. When we felt we were in a good position, Nikolai deployed the drone from the bottom of the trench; I was terrified that the grenade would release prematurely and detonate on take-off, but the deployment mechanism seemed reliable. After we had launched and reliably flown a couple of times, we began to settle into a rhythm. I was keen to do the job for a couple of days to placate the battalion commander, for me to understand the environment, survive and go home rather than worry about the grenade going off in our trench.

It wasn't long before my mind began to drift back to the First World War and the eerie similarities to where I now found myself. More than a century had passed since those trenches were dug, yet here I was, crouching in a damp front-line trench reinforced with wooden slats, pallets laid on the ground to keep our feet out of the filth. I tried to talk to Nikolai over the relentless soundtrack of gunfire, shell bursts, and the distant rumble of armour. The stench was unbearable. My mind felt numb, overwhelmed by a cocktail of fear and adrenaline.

For the first time in my life, I finally understood my grandfather, the look on his face whenever we tried to talk about the World Wars, the pain in his eyes, the sudden irritation and change of mood, the unspoken plea to change the subject. Now I knew why.

I could never explain to anyone back home what it felt like; the complex mix of dread and exhilaration, the disgust every time the wind shifted and carried with it the smell of death. There was no escaping it. All you could do was pray to survive the next few minutes and cling to the hope that everyone would make it home.

I also understood the value of my army training in a way I never had before. The endless repetition of drills, the field-craft, the instincts drilled into me, all of it kept me alive. It taught me to focus, to read the body language of those around me, to recognise the subtle shift in atmosphere that meant trouble was coming. You learned to keep your head down, to predict the rhythm of the shelling by counting between salvos, and to move the instant you had the chance.

But unlike the soldiers of the Great War, we had a third dimension to worry about, the sky. Enemy drones made it necessary to glance upward every few minutes, checking whether the buzzing overhead was friendly or a killer on its way to drop death onto our position. We never stayed in the same place twice. After every deployment, we shifted, always aware that the enemy might be tracking our returning drone to pinpoint our exact location.

I explained to Nikolai that I would do everything in my power to ensure Ukraine received the aid it needed. Still, I admitted my doubt that anyone would fully understand their urgent request for drone equipment. What I didn't tell him was that Damo had already sourced, or been donated, several new drone systems that were en route. There was no point pretending that his influence hadn't affected me, he knew, as I did, that every piece of equipment he asked for would be reported back to Australia, the UK, and Ukraine's European allies.

It was dawning on me that Max had likely already briefed Nikolai and the battalion commander that the Australian Federation of Ukrainian Organisation's was in touch with drone suppliers and manufacturers, outlining their operational needs and even collaborating on new technology. What I was witnessing wasn't just aid, it was the birth of a shared platform for the evolution of battlefield technology, and I was there to see it happening in real time.

Once our drone was launched, we displaced to a slightly different position, ensuring its return would not betray our location, and prepared for the next sortie.

Once in flight, the drone's noise was masked by the battlefield and difficult to see due to the smoke drifting across the trenches. The Russians were using high-calibre artillery, which included the occasional phosphorus shell. The smoke and smell were awful, mixed in with the odour of faeces and urine in the trench. Even a faint hint of phosphorus smoke could scorch your lungs, irritate your eyes and induce gagging. The fact that the Russians openly used phosphorus shells in violation of the Geneva Convention didn't surprise me; they clearly didn't care. Their intent was not only to inflict damage but to intimidate the Ukrainians into submission.

Preparing drone to carry aerial modified grenade

I was terrified but tried to focus on the job at hand; once a drone was launched, it was flown up to four hundred metres high in the direction of the Russian lines. I started by watching Nikolai pilot the drone while I navigated and relayed information. We were over enemy lines within five to six minutes, looking for a target.

Positioning the camera at a ninety-degree angle to the drone, we were aware that we had positioned ourselves over a target. As soon as Nikolai confirmed being on target, he released the grenade within ten to twenty seconds. This was a well-rehearsed ritual for him; I was certain he had performed it countless times previously.

After enduring relentless artillery fire for a while, I sensed being ensnared within a living hell, with no escape from the ceaseless downpour of fire. The days moulded together, with the constant explosions' sound and the smoke's scent merging into an unending facet of my existence. Soon, my nerves were frayed, my mind numbed by the relentless barrage. Every sound triggered jumps, my body tense with fear as I contemplated each moment's finality. Time slipped away, day and night losing distinction. We slept wherever we could somewhere that was dry and warm. Two of us worked with the drone team whilst the others prepared the vans to return to Kyiv with refugees or casualties The sun transformed into a distant memory, concealed behind the thick curtain of drifting smoke and ash shrouding the battlefield.

I wanted to go home, back to Kyiv, but now understood that it was going to take time to prepare the vans to move the wounded or elderly and in-firm.

Ukrainian Artillery Unit overlooking Russian Lines

Debris littered the trenches and ground beneath us, and the air hung heavy with dust and the war's smog. Inhaling the noxious fumes left my throat raw, while occasional phosphorus shells intensified the torment, burning my eyes. Amid the struggle to survive and think, an inexorable force compelled me to persevere. Amid the fear and desperation, glimmers of hope and resilience emerged among the team.

Each day demanded fresh reserves of strength and courage, and I refused to allow the enemy to shatter my spirit or bend my will. For a brief span, I became a warrior of necessity, fighting not for glory but for survival, holding firm until the thunder of artillery finally faded. It was a war of attrition, an endless nightmare, a test of endurance, resolve, and the capacity to withstand fear. Yet I knew I was not alone. Others like me, there were many British, American, Australian aid workers, enduring with the same intensity, clinging to hope in the midst of chaos. Even in that nightmare, the glimmer of a brighter tomorrow never vanished. Still, just a single week beneath the shelling left me shattered and irrevocably changed. I longed to leave, though I carried nothing but admiration for those who stayed, unflinching, to continue the fight.

Surviving the mental torment of a war-zone is an ordeal beyond imagination. It leaves scars that do not fade, especially for those forced to remain. To endure, one must summon untapped reservoirs of resilience and inner strength while navigating relentless chaos and violence. No one can face it alone. Solace comes in the fragile sanctuary of camaraderie among fellow aid workers and soldiers who shoulder the burden together, forging bonds that , transcend the horrors of war. Yet the weight remains. Loss, fear, and memories etched in fire cling long after the battlefield is left behind.

Coping takes many forms, dark humour as a shield against despair, finding meaning in the mission, or drawing strength from the shared duty to protect one another. But the torment lingers. For many, it demands more than time; it requires support, understanding, and the courage to confront wounds that are invisible. Only then can the long journey toward healing, recovery, and some fragile sense of normality begin.

A powerful wave of relief washed over me as I successfully distanced myself. Due to health issues, my time in Ukraine had limitations, and I was enthusiastic about optimising what I had. Over the past months I had lost eight kg's in weight and my health was deteriorating fast. While seeing a liberated city boosted my spirits, I still looked forward to leaving and returning to Lviv. I needed to plan my travel carefully as I was unable to eat and travel. My time in Ukraine was getting more and more limited.

By the time I rejoined the team, they had already modified the vans with rope and netting inside, converting the vans into make shift ambulances and begun loading the wounded and squeezing in as many refugees as we could carry into the vans.

I left Kherson a different person to the one who had arrived only weeks before. The thought of the front-lines and of the men and women enduring life in the trenches, had already begun to invade my nights, surfacing in vivid nightmares that I feared would soon become an enduring problem.

Travelling back to Kyiv took longer than driving to Kherson; the checkpoint guards were much more sensitive about people leaving the area and taking usable materials away from the combat zone. Each stop included a complete inspection of the vehicles and an explanation of the equipment that we had with us, even down to our own sleeping bags and food. It was an understandable requirement, though annoying. Once the checkpoint guards knew that we had completed an aid run and had wounded on board, they were more compassionate and could see we were tired. It was at this point we understood the importance of a local fixer and interpreter; with Nastasiya remaining in Kherson negotiating checkpoints without Max our interpreter would have been impossible.

When we returned to the hostel, our place within the aid community had shifted. We were no longer just volunteers, we were now veterans who had been to the front lines. The others in the hostel wanted to hear our stories, often more than we wanted to tell them.

I found it easier to speak with Max and the others who had shared the aid run with us. They understood without explanation. But talking to newcomers or to the press was off-limits. I simply didn't have the words to explain what we had seen, beyond confirming that we had completed the trip and helped relocate refugees.

Our experiences had earned us quiet respect. We were now part of the hostel's inner circle of veterans, and even the soldiers attached to the battalions began to show a deeper sense of camaraderie toward us.

Volunteer Ambulance Team

It took me at least two days to decompress before I could bring myself to leave my room and begin reintegrating with the others. I kept my head down, barely speaking, still replaying the images of what we had just been through. Max, of course, seemed to sense it. He knew I had driven an ambulance with UK Aid before, and after a few days he came to me with another request; this time to help him cover a shift driving a recovery ambulance, supporting the local emergency crews during air raids in Kyiv.

At first, I resisted. I didn't want to get myself in any deeper. The thought of running toward another crisis so soon after returning from the front lines felt unbearable. But Max was insistent, and somewhere deep down I knew I couldn't just stand aside. These people needed help, and if I had the skills to provide it, then I had no excuse not to. Eventually, I agreed to join him, not just as a driver, but as a standby medic.

Max knew I had just completed a first aid refresher before travelling to Ukraine, and before I fully understood what I'd signed up for, he turned up in a Mercedes long-wheelbase ambulance, almost identical to the ones I'd used before.

At first, we set a reasonable plan: two twelve-hour shifts every three days. It sounded manageable. But war rarely follows a schedule. Before long we were taking more shifts, most of them overnight, grabbing what sleep we could during the day while still coordinating aid deliveries.

By late November, Russia had begun to intensify its air raids as winter approached, hoping to cripple the capital and pressure the population into surrender by freezing them out. Most raids came late at night or in the early hours of the morning, so our shift would usually begin around 10 p.m., just as Max pulled up outside the hostel.

Once on station, we sat quietly in the cab of the ambulance, headlights off, waiting for the phone to ring. I couldn't understand the Ukrainian chatter on the radio, so I relied on Max to translate when we got a call. There was always a strange calm in those minutes of waiting, as though the city itself was holding its breath.

But that calm never lasted. The first wave of sirens always broke the silence like glass echoing through the city being repeated on the phones. We could feel the distant thud of explosions and see the glow of fire on the horizon before we heard them, a low vibration that passed through the ground and into our chest. Then the phone buzzed, and we were off.

Our first call that night came just after midnight. A housing block had been hit halfway down on the fifth floor, devastating the entire south side of the building. By the time we arrived, the place was a ruin, a yawning black hole where homes had been only hours earlier. Flames licked at the exposed rebar, and the air was choked with plaster dust and smoke. The smell of burning insulation and scorched concrete was overwhelming.

Max and I parked the ambulance in the shadow of the next building and ran toward the site with the other responders. The local crews were already at work, but there were not enough of them for a strike of this size. We joined in without hesitation, knowing that every minute mattered. Somewhere under that collapsed floor there were people still alive; and we had to reach them before the smoke, the cold, or the weight of the rubble finished what the missile had started.

I had never felt such urgency. Even in previous air raids, there had been moments of order, procedure, a chain of command that gave some semblance of structure to the chaos. Here, all of that dissolved. There was no time to wait for instructions, no time to think about our own safety.

We went in with bare hands, hauling debris aside, passing chunks of masonry down the human chain forming around the building.

The first voice we heard was faint and muffled, but it electrified us. The crew shouted for silence. We froze, listening. There, a woman's voice, calling weakly for help. We followed it, digging toward her, trying to open a crawl space large enough to reach through.

Recovering a Women in Kyiv, from under rubble

Hours passed like minutes. We found our first survivor, a woman with a crushed leg, just before dawn. Max and I loaded her into the ambulance and sped to the nearest triage point, then returned for more. Every trip back to the ruins felt harder, but we kept going.

The raids were relentless that winter. Even when the anti-air systems intercepted most of the missiles and drones, there was always one or two that got through, creating debris, burning fragments falling across the city, damaging homes, infrastructure, and lives. Sometimes we would arrive at a site to find not one but two or three impacts within the same district, each demanding our attention.

Soon, the rhythm of it became part of us: the late-night phone call, the rush through darkened streets, the acrid smoke, the digging, the loading of the injured, the sound of sirens merging with the sound of crying. By the end of every shift, my body ached. My hands were blistered and raw despite gloves. My clothes grew stiff with dust and blood. And yet, when the next call came, we always went.

The second shift was even busier than the first. The night sky over Kyiv was still faintly red as we arrived. The air smelled of smoke and pulverised concrete, the bitter dust settling on our tongues with every breath. The air raid sirens had only just fallen silent, their wail replaced by a heavy, oppressive quiet, broken now and then by the sound of distant impacts.

The building before us had been an apartment block only hours earlier. Now it was a broken spine of concrete and steel, its floors pancaked, its stairwell ripped open like the pages of a book. There were no neat edges, just twisted rebar, chunks of masonry, and shards of glass glittering in the half-light.

We didn't wait for instructions. Those of us who had arrived ran toward the wreckage, the instinct to do something stronger than the fear that we might still be targeted. Searchlights from emergency vehicles and soldiers' torches cut harsh angles across the ruins, throwing our shadows long and thin over the debris. Somewhere beneath our boots were the people we had come for. We could already hear them.

At first, it was hard to tell if the faint cries were human or just the groaning of the building settling on itself. But then came a distinct sound, a woman's voice calling, muffled but clear enough to stop us in our tracks.

"Допоможіть!" Help.

I dropped to my knees without thinking, brushing away chunks of brick and plaster with my gloves, then discarding them entirely to feel more clearly with my bare hands. The cold concrete bit into my skin and my nails caught on jagged metal, but none of it mattered. We clawed at the rubble like animals, scraping away anything we could shift, tossing it behind us. Every few seconds we would stop and listen, trying to triangulate the voices. They came from deep below, the pitch of the cries sometimes shifting into a hoarse cough as the dust suffocated them.

Max appeared next to me, already bleeding from a cut on his forehead. His expression was grim but focused. Without a word, he began to pry apart a section of shattered plasterboard, revealing a narrow cavity beneath. Together, we widened it, stone by stone, passing each piece back in a frantic rhythm.

The hours blurred together. Time seemed to stretch, an endless cycle of digging, listening, digging again. The sounds around us became part of a terrible orchestra: the crunch of debris underfoot, the rattle of metal bars being bent away, the muffled sobs of survivors now waiting for medical treatment nearby. Somewhere, a baby was crying, thin and sharp, a sound that cut through everything else like a knife. It became our compass. As long as we could hear that child, we knew there was still life to save.

At one point, we uncovered a man's arm. Grey with dust, but still warm to the touch. The rescuers around me went silent. For a moment we all just stared, not knowing if we had found a body, or someone who might still be alive. Then the arm twitched. Just once.

The silence shattered as everyone shouted for more hands, more light.

We worked faster, desperate not to crush him as we pulled away the debris. Every movement became deliberate, every stone shifted with care. The man groaned as we exposed his face, then his chest. His eyes fluttered open, and in that instant, all exhaustion left me. We had found him alive.

The soldiers fashioned a crude stretcher from broken doors and lengths of pipe. When we finally pulled him clear, there was a cheer, not loud, not jubilant, but the tired cheer of people who had been holding their breath too long. He was carried to the waiting medics, his breathing ragged but steady.

We clung to hope, focused on the living. It was the only way to keep going. We dug as fast as we could, listened for the smallest sounds, and prayed that the ruins above wouldn't collapse further. As the night dragged on, more voices were found, some faint, others already fading. Not all could be reached in time. I will never forget the moment we uncovered a woman and realised she was already gone, her hand still clutching the arm of the child beside her, who was miraculously alive. We had to gently prise the child free, the dust sticking to his tear-streaked face. He didn't cry. Just stared at us with wide, uncomprehending eyes.

Dawn came slowly, turning the smoke into a pale orange haze. We were filthy. Our hands were raw and bleeding, our clothes streaked with soot and concrete dust. The stench of burning mixed with the coppery smell of blood and the sour tang of fear-sweat. Around us, the building still loomed like a tomb, but now there were gaps, narrow paths where we had dug. Each one marked hours of labour. Each one was another chance at life..

The last person we pulled out was a young woman, her legs pinned under a collapsed beam. It took nearly an hour to free her, and when we finally lifted the beam with the help of a jack, she screamed, a raw, animal sound that seemed to empty the air. She was alive, but her legs were shattered. The medics sedated her quickly, their movements calm and precise, as if they had done this a hundred times before.

When we finally stepped back from the wreckage, the sun was fully up. The city was stirring around us, buses running again, people heading to work, as though the night's destruction were just another part of life now. We sat in silence, too exhausted to speak, staring at the ruin we had just spent hours inside.

At this point, we had been working together in Kyiv for weeks, recovering the trapped and the wounded after air raids. The work was exhausting and difficult to come to terms with, especially when we were confronted with the grim task of finding bodies among the debris, some of them infants. Over time, we stopped talking as much and simply worked, shift after shift, falling into a rhythm of quiet understanding. Our shared sense of loss bound us together, broken only by moments of euphoria when we managed to pull someone out alive, another survivor.

Later, when journalists arrived and tried to ask what we had seen, I had no words. How do you describe the sound of the living trapped under tons of concrete, the way their voices seemed to echo inside your skull even after you'd stopped digging? How do you explain the feeling of pulling someone from the rubble and knowing you had snatched them back from death, if only for a little while?

All I could do was nod when they asked if we had found survivors. Yes, we had. And I prayed silently that the next time the sirens sounded, we would be there again, ready to dig, ready to listen, ready to bring someone else back to the light.

After a couple of weeks running the ambulance, I had been overcome with exhaustion and needed to take a break. At last, the moment had come to stop, breathe, and allow ourselves a well-earned rest, a brief chance to let the weight of everything we had endured begin to settle.

Whilst Max understood that I needed to move on, and that my time in Ukraine was limited by a visitor's visa; he struggled with my decision to return to Lviv. Over the past several weeks, Max and I had become a team. We had begun to instinctively understand one another, sharing an emphatic connection forged through moments where life itself seemed to pause, watching others pass from life to death, or hover somewhere in between.

I promised him that I would return in the spring of 2023, and in the meantime, I would continue our work from the UK and Australia. Since I had first begun my work in Ukraine with Max, the Australian Embassy team had grown significantly, with nearly a hundred new volunteers. After speaking with Damo, I learned that most of these new recruits were deeply committed to supporting Ukraine. Many were veterans of the Defence Force, experienced aid workers, or federal police, people with the skills, discipline, and resolve to make a meaningful impact on the future of our mission.

Max had become a brother to me, someone I would never forget. Like Andy from UK Aid, he was a person I now knew Instinctively and would continue working with well into the future.

CHAPTER 9

Returning to Lviv

By now I was unwell, tired and feeling totally broken. I rested as much as I could and booked the overnight train again back to Lviv. It was time to take a break and reassess our position summarising what we had achieved and make a plan for the next couple of weeks before travelling home.

The visit had proven to be productive thus far. We had established contact with five small to medium aid agencies and had started comprehending their functioning within the country's military structure. We had also been able to connect a number of drone units to Australian and American Manufacturers.

I had formed connections with UK Aid, Ukraine Patriots, numerous hospitals, Unbroken, a rehab hospital, and a UK-based surgical organisation. The longer I stayed in Ukraine, the more English, American and Australians I encountered. They seemed to be everywhere, often noticeable with their Auscam backpacks or standard-issue kit bags. Whenever we interacted, we talked about the insufficient support from Australia and shared our

disappointment regarding the absence of an Embassy in Ukraine.

As soon as I could get back on the train, I was on my way back to Lviv. By now, the Mayor of Kyiv had requested all non-Ukrainian residents to leave Kyiv as the power cuts were becoming a significant problem; the situation was dire, everyone at the Hostel who could leave was preparing to go, the mood was bleak, it was winter, cold, damp and just dark. The city's residents knew that the colder months of the war were on their way, and it wasn't easy to remain happy. The usual comforts of everyday life had become a luxury, and I was beginning to understand how little someone could survive on and relish the amenities of hot water, light and warmth.

Booking transport online stood out as the most straightforward method, raising eyebrows due to the enduring presence and reliability of the Internet, even during extended power outages. The irony lay in that the Russian forces and the Ukrainian Army relied heavily on the Internet to operate, and it remained consistently accessible when electricity flowed. Ukraine's extensive rail network and the exquisite train station architecture emerged as significant highlights, adding to the nation's charm.

I was looking forward to going back to Kyiv's Central Railway Station and wanted the opportunity to see it during day-light. It is a marvel, with its Neo-Renaissance facade adorned with intricate carvings and a grand clock tower that stands as a city symbol.

Without a translator, the hardest part for me was trying to talk to the booking clerk at the station to find the correct platform and carriage. Many of them did not speak any English, and I began to rely heavily on a language translator on my phone. The stations were hectic and carefully managed by the military, with hundreds of soldiers of all ages moving through mixed in with civilians. Just getting to the correct platform in time was the main priority. The past months had taken their toll, and I slept through most of the journey back to Lviv, so relieved to be leaving that I could barely remember parts of the trip at all.

As soon as I got back to Lviv I booked myself into the nearest hotel. The next morning, I left the Central Park Hostel as soon as I could after a quick breakfast of dry bread and tea. The damp bedding and smell made me feel sick, and the mould on the walls made me cough; out of all the places I had stayed in so far, this had been the worst. At least most of the hostels had been dry and warm.

As planned, I called a taxi to take me to the Catholic College near Stryiskyi Park in the south of the city and headed off as quickly as possible. I was looking forward to a few days' rest to gather my thoughts and plan next week's movements.

I reached the College within twenty minutes and proceeded to the accommodation block reception to check in. A contact from our group had advised me that ample lodging would be available. However, upon arrival, I discovered a conference was taking place, and all the rooms had already been occupied. Surprisingly, the individual we had entrusted to arrange our accommodation had yet to coordinate with the College and had made unwarranted assumptions.

Unfortunately, in my initial inquiries, I had mentioned our association with the Australian-Ukrainian embassy. This information had been unknowingly conveyed to a team operating on the campus from private offices, Smart Medicinal Aid. As a result, a well-dressed, younger, middle-aged woman approached me to ascertain my identity and purpose.

Smart Medicinal Aid

As one of our main goals was to find local Ukrainian Aid agencies, I was keen to talk to her to find out where Australia could support the agency. Irena invited me to the Smart Medicinal Aid office to discuss why I was in Ukraine. Like many times before, I mentioned that we were working with the Australian Ukraine Embassy providing her with my references and the business contact of Volodymyr Shalkivsky, the Australian Ukrainian Embassy's Chargé d'Affaires. She seemed initially hostile and reluctant to verify my credentials, saying that she knew the Ambassador as a personal friend and had never heard of Shalkivsky, his aide, stating that she felt that my credentials might be fake in front of everyone in the office. Her attitude surprised me; her behaviour was unusual as I had already contacted Volodymyr via a messaging service where he was waiting for her call. When this was communicated to her by email and face-to-face, however, she remained reluctant to verify my details.

It was then that I began to sense something was seriously wrong. Her attitude had shifted; there was a heaviness about her, a quiet trouble that clung to the air. Those around her betrayed their unease through nervous gestures and disjointed speech. Having spent so much time in Kyiv with legitimate aid agencies under similar circumstances, I instinctively knew something was amiss, but I needed to understand what it was, and whether there was still a chance to set things right.

Still, she insisted that I stayed locally, offering to provide me with accommodation while she confirmed my details.

This was the first time that my credentials has been questioned and her attitude had unsettled me deeply. I was seriously concerned that she had refused to call the Embassy's Chargé d'Affaires, Volodymyr. Why would she act this way?

During the morning, Irena asked her colleague Yermani to take me to their store room to meet the team. Irena insisted that I leave my luggage in their offices at the College, and Yermani and I travelled to the Smart Medicinal aid Stores at a Unit in Staroznesenska St, Lviv Oblast, I initially thought this could have been a breakthrough and felt some encouragement from the offer of a tour; however, I was only going to regret this later.

On arrival at their warehouse, we parked at the front of the complex and made our way around to a large storage area at the back. An ambulance was parked there, with five people standing nearby. Yermani introduced me to those closest to the vehicle inviting me to take a photograph of the ambulance.

Moments later, a man appeared on our right. He immediately began demanding to know who I was and why I was there. Bruce was an imposing figure, tall, broad-shouldered, and physically intimidating, with a face like a "bag of spanners," brutal and hard. As he emerged from the office beside the storage area, I noticed him casually tucking a "Glock pistol" into the back of his trousers.

Among the paperwork and boxes on a desk, another pistol lay. This sudden development made me extremely anxious. I conversed with the man, who introduced himself as Bruce, attempting to calm him down. I explained that I had been invited to the storage area by Irena. Bruce argued with Yermani, questioning why I had been brought to the warehouse. The combination of Irena's attitude and Bruce's reaction had completely unsettled me. The situation was spiralling out of control, raising more questions than answers. Why were guns so readily available within an aid organisation; and why were people behaving with such open aggression?

At this point, another man called Liam stepped in, offering to provided me with a cup of tea in the canteen. I was understandably keen to get away from Bruce, an imposing aggressive individual, I immediately felt at risk, he was a brute. We quickly departed from the storage area walking back to the canteen to have a pot of tea, taking a deep breath to calm down.

As Liam and I left to walk to the canteen, I could hear Yermani instructing the remaining team members on how to remove items from the ambulance and swap them with older and cheaper items from another two ambulances parked out the front of the complex. I could understand this conversation, as both Liam and Bruce were English speaking.

I was pleased to see that Liam was a more affable character, someone I felt at ease with, especially when he brought out a pot of tea in the canteen and invited me to sit down. We shared a brief conversation about Ukraine and the various aid organisations, during which he appeared more settled and willing to accept my credentials. The moment was short-lived, however, as Bruce soon joined us and, once again, his manner turned aggressive.

I found Bruce's comments and attitude very odd under the circumstances. I was visiting Smart Medicinal Aid by chance and had been invited to view the warehouse by Irena, the CEO. He would not accept that I was a Chaplain visiting the Catholic University for accommodation, even though I had shown him my ID and full credentials. He just appeared to be paranoid, something that I had not yet encountered in this type of situation, making me feel very uncomfortable and even more suspicious.

We returned to the College after a short break. When we returned to the Smart Medicinal Aid offices at the College, I was given directions to the food hall and ate lunch before returning.

Upon returning to the offices, I spent the afternoon in quiet observation. I watched the young individuals in the office as they worked alongside Irena. During this period, I also observed that my luggage had been opened while I had been in the storage area. Most of the conversation occurred in Ukrainian, a language I did not comprehend. The atmosphere in the office had been uncomfortable amongst the staff, who were looking at me from time to time with unsure glimpses.

At the end of the day, Irena insisted that I stay in a Smart Medicinal accommodation block located behind the University while she contacted the Australian Ambassador to clarify the situation. Although I was confident the problems could be resolved, I had grown increasingly uneasy about her and the team's behaviour. To keep my options open, I quietly booked a couple of alternatives for the following day and resolved to leave the accommodation at first light, slipping away before anyone expected me to appear. Something was wrong, I just couldn't understand what it was or why.

Yermani drove me in silence to a shared accommodation block behind the University, deflecting most of my questions with a polite claim that he didn't understand me, though I knew he spoke perfect English. I was allocated a room shared with two other families; the apartment had several bedrooms, along with a large communal kitchen and bathroom. It was late, and I felt both confused and disoriented. I decided to stay the night but planned to leave at first light, quietly passing a thank-you message to Tetyana, a woman I had just been introduced to before making my way back to the Dream Hostel in Lviv. I resolved to avoid any further confrontation and sort out the misunderstanding through the Embassy later.

Tetyana and her daughter were kind enough to offer me a small meal and a glass of wine before bed. I left most of my equipment ready to go but felt so uneasy about the situation that I barricaded the bedroom door and set my alarm for five a.m., well before the scheduled eight o'clock pickup.

The Assault

As usual, I woke early and slipped into my routine. After the reaction from the Smart Medicinal team the day before, I was determined to get moving. I packed away my sleeping bag before even considering breakfast. Sharing a bathroom with others in the apartment, I tried to keep my disruption to a minimum.

As agreed, I left the room keys on the kitchen table along with a letter with Tetyana for Irena. To my surprise, she was already awake, perched at the table working on her laptop. I thanked her politely for her hospitality, then slipped away as quietly as possible, avoiding any fuss.

Whilst I had booked a bus to Warsaw, I initially intended to move back to the Dream Hostel in Lviv as I felt comfortable there, knowing the people and familiar with the surroundings. I had more people to meet and had plans to catch up with my contacts in the area. I booked a taxi via a phone app and gathered my belongings to wait outside, pulling my bag down the long stairs around the side of the University dorms. Moving down the slope was no more manageable than it had been coming up, as my bags were heavy, and the wheeled trolley kept unbalancing.

I was a good forty metres down the stairs from the door towards the end of the building as I paused for breath and rearranged the grip on my bags. I got to within ten metres of the end of the building and became aware of two men moving towards me from my right-hand side; they had just exited a car parked on the street just up from the entrance and moved quickly towards me. It was difficult to react in time as my bags limited my movement.

Before I realised what was happening, they had stunned me with a flash of bright light and kicked me in the side. The shorter of the two men cuffed my right arm and forced me to the ground. The pavement was cold and hard. I felt disoriented and stunned by the speed of the attack. As the taller of the two spoke, I grasped it was the two guys from the previous day, Liam and Bruce. They were quick and efficient with what they wanted to do; Liam knelt on my back, keeping me as still as possible, whilst Bruce stamped a few times on my right arm. Liam then knelt down using what looked like a medical scalpel to just cut the skin slightly on my index finger while Bruce whispered in my ear, "Fuck off and don't come back; if you tell anybody what you've seen, we will kill you and cut off your fingers". The whole episode only lasted thirty seconds, and it was over.

The two of them left me on the ground stunned, strolled back to their car, and drove off. l lay on the pavement for a while, reeling in pain, trying to figure out what had happened and, more importantly, why. What had I seen at Smart Medicinal aid that I shouldn't have, and more importantly, what would I do now? My arm was in agony, and I quickly understood that it would be a problem as it started to colour up within minutes; it was difficult to move my thumb and fingers, and the pain moved up my arm when I tried to move.

The first aid kit was within easy access, and I had some old morphine sulphate tablets, so I took one and gathered my thoughts. These guys were serious players and well-connected, whilst Liam had seemed the most reasonable one the day before. Bruce was the easiest to read, a real hard man ready to do whatever it took to get the job done. Mind you, the calm coolness with which Liam had performed made me feel even more uncomfortable; he was a cool operator ready to buy you a cup of tea one minute and slit your throat the next; either way, they were paid professionals.

It then began to dawn on me that they had taken me down to the warehouse the day before not for a tour of their Ukraine Aid facilities but to meet these guys as they wanted to know more about me and had convinced me to leave my bags at their offices so that they could go through them at their leisure. By offering to put me in accommodation for the night, they had also set me up ready for the following day. They knew who I was, why I had turned up at the Catholic College, and that we had come across them just by chance, but something had spooked them into reacting this way. The organisation was not what it seemed to be on the surface, and there had been something that I had said or seen that worried them. Either way, I needed to move on and get medical treatment for my arm; they had deliberately wanted me out of action for a reason.

By the time the taxi arrived, I was back on my feet and beginning to think tactically; I was too old for this shit and needed to get out to Poland, where I could get some treatment. So, I changed my plans and did what they wanted me to do: go dark and disappear.

The taxi driver was surprised to find me clutching my arm and reacted by kindly loading my bags into the taxi before I passed him a copy of the address of where I wanted to go. Trying to think through the situation, I had found myself in, I had no idea how well, Irena and Smart Medicinal aid were connected locally which troubled me deeply and needed to get out of the country as quickly and quietly as possible. I left a message for Damo and the Australian team by phone then took the SIM out of my phone removing the battery, to ensure that I had become as invisible as possible.

I made my way to the coach station and, with the help of someone who spoke English, managed to find a bus. It had been delayed, and the driver was already irritated, annoyed that I couldn't properly pronounce "Warsaw" or load my own bags due to my injury. I tried to explain that I couldn't lift them, but he just flapped his arms around in frustration, complaining loudly in Ukrainian.

I stayed calm and asked for help again. Thankfully, someone nearby who spoke some English had overheard and stepped in to explain the situation. The driver begrudgingly loaded my bags. It was clear that it was going to be a long, painful day.

The young woman who helped me was also heading across the border into Poland to meet her Irish boyfriend. She'd been learning as much English as possible, and people like her were absolute gold when you found them. I spoke with her quickly, and she agreed to help me navigate the trip.

The bus eventually departed, but tensions were already high. Everyone was frustrated, we had been informed there was a problem and that we'd have to change buses multiple times during the journey due to another coach breaking down. I wouldn't have known any of this if my new friend hadn't translated.

The disruption, though, suited me. I had deliberately chosen a different route out of the country given my situation. The more erratic and unpredictable the journey, the better. My new plan was to arrive at the border later in the day, crossing at a quieter location on foot before switching to another bus on the other side.

By now, I was completely paranoid after the final encounter with Smart Medicinal Aid. I needed to take every precaution to avoid being intercepted by another one of their team, especially someone like Liam or Bruce.

As planned, we changed coaches once before we got to the border and were delayed. It was early morning when we arrived when it should have only been a ninety-minute trip. The border crossing was quiet, which I later learned was a bad thing as it gave the border guards a better opportunity to isolate each person, tip out their bags, and search the bus from top to bottom. If it had been busier, they would have been under more pressure to work faster and would have cleared us through more quickly. Again, for some reason, the border guards were great with me. Again, they displayed a favourable disposition towards the English, American and Australians. They examined me from head to toe, stamped my passport, and swiftly motioned for me to proceed. However, it's worth noting that checking everyone and meticulously searching the bus took four and a half hours. With only a limited supply of water and having finished a single protein bar, I was beginning to experience dehydration, and my foremost desire was to continue onward. The Ukrainian border guards spent hours stripping the coach looking for hidden compartments hiding contraband and men trying to escape conscription. We waited hours before a dog team arrived to double check the coach.

While the guards were distracted searching the coach, I slipped away and made my way to the nearest footbridge. Under the cover of darkness, I crossed the border on foot and switched to a different route on the Polish side.

At first, I was anxious about how the Polish guards might react to me arriving alone and on foot, especially after deviating from the planned route. But to my relief, they were surprisingly friendly and supportive. Within minutes, they had waved me through and I was moving again.

I boarded the next coach, the mood amongst the new group was depressed and quiet, everyone was tired and cold. We sat for hours on the couch without stopping and arrived in Warsaw over six hours later. I took an expensive taxi from the coach station to the hotel and almost passed out from fatigue.

I was just grateful to be alive, and safe.

The pain in my arm woke me up early in the morning; I took a fist full of painkillers with a protein bar before getting out of bed and getting on with the day. Looking out of the hotel window, it was difficult to tell what time of day it was as the whole city was covered with a shroud of mist blended in with a glimmer of lights from distant buildings. From the tenth floor of the hotel, I had a great view of the city centre and the imposing structure of the Palace of Culture and Science. Large highways crisscrossed the city with yellow trams moving up and down the middle of the roads in an organised dance that mesmerised me for a while. I sat and watched the morning in a daze and tried to unravel the events of the last couple of months, trying to understand what had happened and why.

Every time I travelled abroad, I hide a moderate amount of American cash in a money belt in case of emergencies. In this particular situation, I found myself in need of determining the severity of my arm injury. Given the circumstances, I chose not to go to a hospital formally because I had become paranoid and had concerns about whom I could trust. Smart Medicinal Aid had significant connections, and I wanted to avoid causing a political incident. After hours of searching and help from the hotel, I located a small radio-logical clinic that spoke English and was willing to perform an imaging of my arm in exchange for cash without any official records. It had been surprising how easy it had been able to arrange. The X-ray simply confirmed a minor hairline fracture and heavy bruising, particularly where the hand cuffs had been. My arm and hand throbbed with pain and had numb spots. In any case, I improvised a basic strapping to support it and walked back to the hotel.

Trying to arrange an urgent flight home from Poland was challenging and took most of the morning. I was scared, tired and sick and just wanted to go home.

As soon as I had replaced my phone SIM, it didn't take long for the Embassy network to become aware of my situation and what had happened. Reflecting on my earlier contact with Vasily in Lviv, I knew he would have tracked my movements back to Poland and understood that I was now safe, something I had been counting on since our first conversation.

Meanwhile, others in our WhatsApp and Signal groups had also started to grasp the difficulties of working in the region. Although I had been reluctant to share the full details with him, Max and Damo verified Finn's identity and reassured me it was safe to meet. With their encouragement, I arranged to meet him in the hotel foyer. Even though it was late, he arrived as soon as he could, eager to dig deeper into what had happened to me.

Finn, The Film Maker

Finn was everything I had imagined a documentary filmmaker from Ireland to be. He had been investigating Smart Medicinal Aid in Lviv for months, so it was understandable that, after hearing what had happened to me, he was eager to interview me. My only initial concern was that my situation had been confidential, and I was surprised that he had managed to find out about it at all. However, by this point, he clearly knew more about Smart Medicinal Aid in Lviv than I did, and I needed to know more about the risks involved.

What he told me was important; Smart Medicinal Aid was allegedly under investigation for money laundering, and the partner of the CEO, Irena, was a businessman with contacts in the Netherlands and Lviv, including the Splitz nightclub based in central Lviv. A known centre for organised crime and money laundering.

Finn, bubbled with energy, he was indeed an intrepid member of the press, a typical Irishman of towering stature and boundless enthusiasm, had a commanding height and bushy beard that made him stand out in any crowd, his unshaven countenance added a touch of rugged determination to his demeanour. His sharp, focused eyes were always scanning for the next big story, their intensity matching his dedication to his journalistic pursuits.

This outgoing Irish filmmaker was a force to be reckoned with, his passion for storytelling radiating through every word he spoke. A glance at his website and background revealed a man who thrived at the heart of the action, tirelessly chasing leads, interviewing sources, and piecing together narratives with relentless focus. There was always that confident, knowing smile on his face, as if he had already uncovered something significant. Whether pursuing a breaking news story or digging deep into investigative reporting, his unwavering commitment to exposing the truth was a clear testament to both his film-making skill and journalistic integrity.

His presence in the field was as striking as his commitment to the story at hand.

Finn had been keen to meet me and I needed some help, even if it was a splint for my arm. I just needed to try and start to understand what I had become caught up in, so we chatted for hours about the narrative and issue of organised crime in Ukraine.

He was genuinely working on a documentary exploring the stark contrast between honest aid workers, unpaid volunteers driven purely by a sense of humanity and the organised crime networks that had infiltrated certain aid agencies, some operating with the tacit support of the state. His mission was clear; to expose the criminals hiding in plain sight, whether or not the Ukrainian government was aware of their presence. What had happened to me fit seamlessly into the story he was determined to tell.

He told me that there had been a lot of previous employees of Smart Medicinal Aid who had left the company complaining about conditions in the organisation, issues with financing that they had come across, and misappropriation of aid. Whilst they had talked about it with Finn, they had been too frightened to make a formal report, having been seriously intimidated and threatened by the mercenaries within the group, like Bruce and Liam.

After listening to Finn for an hour, I began to understand why Smart Medicinal Aid had reacted so strongly to my presence. My chance encounter had likely triggered a defensive response, they may have thought I had been planted to spy on them by a rival gang, the Embassy or even by the security services. In many ways, I was lucky to be alive. Yet I still didn't know why I had been assaulted and threatened.

What had I seen? The truth was, I was simply exhausted and ready to head home the next day.

I agreed to stay in touch with Finn and work with him to understand why they had reacted the way they did. We both needed to dig deeper into the organisation and go through the photographs I had taken. While Finn's visit had reassured me to some degree, it had also left me deeply unsettled with the growing sense that I had stumbled onto something I might one day regret uncovering.

I went to bed that night with a strange sense of déjà vu, back near the place where my journey in Ukraine had begun, yet now completely overwhelmed. My head spun with the weight of the past months, every moment replaying alongside the dull, persistent pain of my injuries. I swallowed enough painkillers to take the edge off and finally drifted into sleep, just in time to be ready for an early departure.

When I woke, I packed my bags methodically, my mind still on edge. I stayed deliberately within sight of CCTV cameras and kept to public spaces, always alert, always conscious of the need to remain visible and safe.

When I arrived at the airport, I deliberately chose to sit in areas clearly covered by visible security cameras. With my arm in a sling, I felt vulnerable, but also relieved to finally be heading home. The trip had been productive, and I kept reminding myself of the meaningful connections we'd made with aid organisations and Australian manufacturers alike. Thanks to those efforts, more humanitarian aid would soon reach Ukraine safely, and our tech partners now had a deeper understanding of what Ukraine truly needed to defend themselves.

Still, on a personal level, I was beginning to wrestle with the weight of what I'd seen and experienced. This war, and Russia's invasion of Ukraine was unlike any other. The Ukrainian people were fighting with unimaginable courage against one of the largest armies in the world, often outnumbered four to one, not just for their own freedom, but for the principles of democracy and human rights everywhere.

I couldn't shake the belief that the West could not afford to let Russia win. But I was also keenly aware that each act of support from NATO countries inches us closer to the risk of all-out war. In many ways, Russia was already fighting NATO by proxy. With every passing week, more nations and breakaway republics were choosing sides, aligning either with Europe and NATO or drifting toward the growing authoritarian bloc of Russia, China, and North Korea.

Meanwhile, back in Australia and the West, as the news became more confronting, more people seemed to be tuning away or turning off the TV. They sat in comfort, switching off their screens when the news was too confronting. I began to wonder; what would the average person be willing to do, for those who've lost everything, or for the ones still standing, still fighting, on behalf of all of us who value democracy?

Rehabilitation

Living in Australia is fantastic; it's one of the most advanced democratic countries in the world. However, one of the biggest disadvantages of living there is its isolation from the rest of the planet. Understandably, the flight back home proved uncomfortable and long. I had requested assisted boarding, and I felt embarrassed as I sat in a wheelchair, unable to carry my luggage. I struggled to find a comfortable position or get any sleep. The pain in my arm was excruciating, and the injury to my abdomen made me feel nauseous. I had now been on the move for three days with little rest, constantly looking over my shoulder, overwhelmed by fear and exhaustion.

By the time I got home, I was just relieved to see my wife Michelle; I could see that she was upset to see me in such a poor condition in a wheelchair with my arm in a sling. We drove straight to my local hospital for assessment and treatment.

We sat in Accident and Emergency for an hour before being seen, during which time I had just about passed out from exhaustion. Once the nurses realised that I was in a rough condition, they reacted quickly but found it difficult to believe that I had just got back from Ukraine. Most of the clinicians were asking me about my delirium; it wasn't until a consultant stepped in and recognised the handcuff marks on my arm that the team began to understand the gravity of my situation. The visit confirmed a damaged right arm and bruising of the abdomen, where I had been kicked to the ground together with total exhaustion. I needed to rest and sleep before I began to feel functional again.

I was keen to put the whole experience behind me and move on. Still, I found it difficult as Finn and all of the other members of our Australian team, now known as the AKU (Australians and KiWi's in Ukraine), also wanted to understand what had happened at Smart Medicinal Aid. All I could do was to provide them with the information that had been given to me and research the background of the organisation myself.

It took me a couple of weeks to readjust to life back in Australia. During this time, I focused on recovering from my injuries and unknowingly adopted a more reserved demeanour. Naturally, my wife Michelle detected the shifts in my outlook. I frequently found myself lost in thought, instantly transported back to Kyiv and the individuals I had encountered there. I could be in the middle of a conversation or having lunch, but then I would suddenly drift away, my gaze distant and unfocused, seemingly unaware of those around me in the room. My mental health had deteriorated.

Thankfully, my family and GP understood my predicament once I could understand it myself and explain it to them. They helped me adjust by accessing PTSD treatment. I found it challenging to explain to people what I had done, where I had been and what had happened. Most people I talked to couldn't understand what I was trying to explain. Why would you go to Ukraine and what was the Bucha Massacre?

I did want to continue to help Ukraine, support the people and explain to the local community in Australia why we needed to be more involved in Europe and the war against Russia. Initially, most people I spoke to didn't understand, were surprised or just overwhelmed with the information.

Whilst I had to return slightly sooner than planned, I still had a ninety-day theoretical commitment to the team in Kyiv and wanted to fulfill it, so I continued to rewrite training programs for the civilians. I mainly worked on PowerPoint presentations in Ukrainian and English covering recovering firearms, making them safe, the hazards of minefields and navigating the issues of moving around a conflict zone, and submitted them via email. While I was on my visit, we had encountered numerous children and adolescents, especially in rural regions, who had stumbled upon concealed weapon caches in wooded areas. Surprisingly, they would casually retrieve these weapons, nonchalantly draping them over their shoulders as a routine occurrence. It had become quite common to see children carrying assault rifles. Naturally, the local adults and schools were understandably concerned about the safety implications and approached me to develop training materials and presentations aimed at educating their youth.

I expanded the material that I had already written, focusing on weapons handling, escape and evasion dealing with hostile checkpoints and minefields. The most challenging subject to tackle was being taken hostage. Most of the people that I had met understood that the Russians were not taking many prisoners of war but rather shooting them onsite. Trying to convince a teenager to de-escalate a situation if they were taken prisoner was challenging; it was essential to try and make them understand that the quieter and less aggressive they were, the more likely they were to survive. I rewrote the PowerPoint presentations a number of times and even presented directly to youth groups via Skype; once you had made a connection with a group in Ukraine and they trusted you, they were grateful for your support and advice, particularly your technical knowledge of minefields and survival.

In the Kyiv Hostel, I had managed to connect with another representative of the 112 Brigade who provided me with a template of service to cover the ninety days, giving me instructions via email on how to fill it in and create a record, even though it seemed that each battalion made their own solutions, I did what was requested and completed the commitment either way. Whilst you will get an official response to any question that you ask from the Australian Ukraine Embassy, the way in which they operate on the ground in Ukraine is totally different and people in Australia struggle to understand the reality of the situation when you try and explain it to them. The only way to understand what is happening in the region is to go and find out for yourself, then you might be able to make a difference.

Meanwhile, Damo and I had developed a number of business plans in the hope of raising enough funds to go back together, he remained utterly dedicated to gathering sufficient funds for the trip. Recognising the necessity for an Australian distribution point in Lviv, he tirelessly collaborated with the Ukrainian community in Australia to raise finances. While I shared Damo's commitment and enthusiasm for the cause, he frequently called me, some times up to seven or eight times a day; becoming persistent with crucial figures in influential positions within the federation and Ukrainian government. His most incredible skill lay in his ability to communicate persuasively, and he achieved substantial progress by working through the Ukraine Embassy. He successfully convinced them to organise a public flag-signing ceremony with dignitaries while I carried on writing a business plan.

In December 2022, we began to support and develop a couple of volunteers who had volunteered with the AKU become embedded within Ukraine Patriots, Kyiv. Ned a veteran from the Australian Defence Force and Ella, a registered nurse from New Zealand, had directly become involved in the conflict in Bakhmut witnessing the severity of the conflict facing extraordinarily high casualty rates within the units they supported. Operating as medics and military medical aides, they were dedicated to preserving lives and conducting themselves professionally. As our rapport with them strengthened, I could discern the toll the attrition rate and mounting casualties took on them.

All I could do was hold onto hope for their safety and well-being, supporting them however I could. Some of the worst days of fighting saw a platoon of thirty men go out on patrol, with only seven returning, the attrition rate was almost impossible to comprehend. Over the weeks they spent working there, I could see their mental health unravelling and felt deeply for their suffering.

Ella was the first to break. After a heated argument within her unit, tensions boiled over when a delirious casualty lashed out and struck her, triggering a violent confrontation. Weapons were drawn in anger, and for a moment it seemed as if a single wrong move could spark an international incident. That was the moment it became clear: they needed to be pulled out, given rest, and a chance to recover before the situation turned truly tragic

Later, I learned that Ella had pre-existing health issues herself and should never have been assigned to a front-line unit. The Ukrainians were also aware of her background but chose to overlook it because their need for support outweighed their concerns. It just becomes a fact of life that people become a disposable commodity during a time of war.

It was also becoming evident that we were losing younger people working on the front-lines in aid units, where they relied on Google Maps to navigate. Many were only in their twenties with limited life experience and no military or field-craft skills. Relying on Google Maps had become one of the primary sources of loss in this area, where they would get lost or drive straight into the enemy lines. Navigation and field-craft skills were imperative; you needed to know north from the south at a glance, understand where you were and how long you expected to be on the road before the next checkpoint.

Approaching the New Year, I was enthusiastic about using the momentum we had. I dedicated weeks to coding a fresh website and formally began to establish a new NGO with Damo under our name, complete with a standard constitution and financial management protocols. After conducting our inaugural committee meeting and completing the company registration process, we tested the website to make it ready to test online transactions. Although we were aware of a ninety-day waiting period required by PayPal to complete the financial setup of the company due to enforced regulations. Damo remained actively engaged in recruiting new AKU members through the embassy, hesitant to distribute the workload. At this point, I can't claim to have been concerned because I was also very busy setting up the website to receive donations and finalising the company paperwork with him.

The company's development progressed smoothly, and numerous individuals from the AFUO (Australian Federation Ukraine) and the community started enrolling as registered partners with us. As our presence became more evident to others, we formulated a formal business plan known as "Project Gum Tree". The objective was to secure a minimum of eighty thousand dollars to acquire vehicles and establish a compact Australian distribution hub in Lviv. It was a simple plan that provided us with several crucial aspects to improve our safety in the field, learning from the mistakes of my first trip. This included having independent transport and accommodation. Due to the power outages, we later added the need for a small generator and "Star-link" internet dish. Communication in Ukraine is challenging at the best of times and we would need a reliable independent connection back to Australia, UK and America.

During this period, Damo was good enough to present to the Rotary Club in Wagga. He was good at this type of challenge, and I had every confidence in him. We both had strengths and weaknesses and worked well as a team. Damo was the kind of man that could sell ice to the Eskimo's (the Inuit people) or make you believe white was black.

Numerous companies and organisations were considering collaboration with us, which propelled our daily progress. The only concern that began to trouble me was Damo's unwavering determination; he refused to yield and initiated daily harassment of many potential donors to the point of damaging relationships. On one particular occasion, he even levelled accusations against a senior AFUO member, asserting that she lacked sufficient concern about aid loss at the border. He even insinuated her involvement in fraud, hinting at the possibility of an investigation. While I hesitated to completely align with his actions, he organised a group online meeting with several organisation members. It was at this point I began to worry about our relationship.

Damo was keen to be introduced fully to Lana from Ukraine Patriots in Kyiv as the founding director of the AKU, whilst I had initially been reluctant to do it as he tended to dominate people and situations. I was concerned that he would overwhelm her, unable to stop himself from pushing his own agenda; I couldn't put it off any longer and put them in touch directly, stepping back as this had always been the plan. I had started volunteer work to introduce and connect people and organisations. The long term plan for me now was to support, direct and attempt to track the welfare of the volunteers involved with short periods of deployment in Ukraine.

The team grew to over one hundred people expressing interest, I raised the issue of corporate liability with Damo. I had become increasingly concerned that he was recruiting volunteers to work with the team in Ukraine but leaving them to operate entirely at their own risk. I believed that we had a duty of care to those volunteers, but Damo dismissed my concerns, calling it an operational matter. He insisted that each volunteer would sign a disclaimer and a non-disclosure agreement. When I pressed him further, asking whether such disclaimers had even been distributed, he became noticeably unsettled.

From my perspective, the team needed more time to prepare, time to create a secure environment for everyone volunteering through the Embassy network. While Damo had proven to be an excellent salesman and had built valuable political relationships, he lacked operational experience at the level he now occupied as CEO of Funds for Community Aid Ltd. He could not see the broader risks he was taking.

Complicating matters further, we were constrained by limited funding and the ninety-day cooling-off period imposed by PayPal. That period became both a frustration and an opportunity: it gave us just enough time to finalise the company's operating procedures before deploying any personnel. In the hope that we could resolve the issues we faced, I carried on developing the business plan to include financial governance.

We were entering the New Year, and everyone was keen to return to support Lana and Ukraine Patriots during the most challenging season of the year; winter.

A few days later, I received the exciting news that we had been given a substantial cash donation of five thousand dollars from the Wagga Rotary Club in late December, a truly fantastic development. This represented a major breakthrough in our efforts and gave us the momentum we needed to move the company into full operation.

Rather than depositing the funds into PayPal, with its restrictive timelines, Damo chose to open an NGO bank account with the Bendigo Bank and arrange an indirect transfer of the funds. Those few days felt like a turning point: we could finally begin serious planning for the future and start preparing the Land Rover model we intended to acquire.

Despite the disagreements we had faced on other matters, both Damo and I were encouraged by this donation. It was the first tangible sign that our plans were coming to life and that the work we had put in was beginning to bear fruit.

After learning about our progress, Ukraine Patriots and Lana expressed strong enthusiasm for us to visit, which prompted us to explore possible travel dates and supplement the donation with some of our own personal funds. At the same time, other AKU team members were preparing to travel to Poland, joining Ben the newest member of the team, who had recently arrived in Warsaw.

These developments led to detailed discussions with Lana and the formation of a plan to return to Warsaw in mid-January. By that time, our goal was to have finalised the company's structure, completed our operational procedures, and secured medical coverage in Poland for the entire team together with an emergency evacuation plan.

CHAPTER 10

The Betrayal

January 2023 had barely begun when, out of the blue, Damo called me one morning from his car, excitedly announcing that he had just bought tickets to return to Poland in a few days to meet Ben, our newest associate, in Warsaw. I was stunned, we simply weren't ready. I reminded Damo that he could not spend the donated money without my co-signature as a director and that, legally, we weren't yet in a position to spend company funds as we were not tax-ready.

Damo, however, was unstoppable, almost like an over-excited puppy, he wouldn't listen. The conversation quickly escalated into an argument. From his perspective, the urgency was undeniable; Ukraine Patriots had offered us transport in Kyiv, and he believed we had been asked to help immediately.

In hindsight, this moment marked the beginning of the end of our association. Damo had become a casualty of the war, unable to control his behaviour or view the issues with perspective. It set in motion a series of serious problems, some of which I would not uncover until months later.

I began to realise that Damo no longer felt he needed my involvement. Ben had already stepped into my role in Warsaw as a stand-in teammate, and Lana had secured a vehicle for our use. Even so, I tried to impress upon Damo that we still needed time; that we should wait until later in the month to ensure we were fully prepared. Our company's structure and bank accounts still needed to meet legal and tax compliance requirements before any deployment or expenditure.

This became our first real moment of serious, aggressive disagreement. It quickly became clear that Damo was unwilling to listen; he simply ignored my perspective and my explanation of the risks. Despite my strong objections, he departed for Poland on the fifteenth of January, taking all the available funds with him, doing so without proper authority.

He had utilised all the accessible funds for his flights and additional baggage, even going so far as to share details on social media to promote the adventure and gain more likes on his Facebook profile. The situation was truly incredible. Unsurprisingly, our disagreement escalated into another severe argument to the point where he threatened me not to tell anyone else about the company not being compliant.

Shortly after Damo arrived in Warsaw, I began receiving complaints from other associates within the AKU and beyond. Damo's relentless push for "ethical" funding had previously earned respect, but his sudden departure for Warsaw, taking all the funds with him, left many questioning his actions and unethical conduct.

Everyone seemed aware of the situation except Damo himself, who appeared blinded by his determination to project an image of helping others. As the backlash grew, I began receiving threats directed at me over our company's behaviour.

Faced with this untenable situation, I had no choice but to resign as a Director of the AKU and as a registered director of Funds for Community Aid Ltd. In a single night, months of hard work and careful planning by both of us seemed to unravel completely leaving me isolated, frustrated and distressed.

It was at this moment that I realised we were in serious trouble. Damo's relentless determination to secure funds for our return to Ukraine had helped us assemble a sizeable group of AKU volunteers, including retired senior Defence Force and Federal Service members, but we could not afford even the perception of unethical behaviour.

My stomach churned with distress. I admired Damo's unwavering dedication, but I knew, deep down, that he had crossed a line. I had to distance myself from his actions, especially as he continued accepting donations, including from colleagues and members of the AFUO, in ways that could be perceived as fraudulent.

Damo was fully aware of the company's position, financial status, and constitution regulating the use of donated funds. Yet he bypassed all oversight, spending most of the available funds and even asking other team members to contribute blindly to a company that was not yet fully tax-compliant.

In the meantime, time, Damo had started his work with Ukraine Patriots. He visited a lot of remote locations during his visit, completing regular aid runs, including Izium in the Kharkiv Oblast, dropping supplies to the elderly and isolated communities, and engaging with some of the youth groups across the Donetsk Oblast to instil the founding principles of both the AKU and Ukraine Patriots, connecting to community, developing hope for the future, creating light in the darkest of times.

I kept following his work, reflecting on the immense challenges he faced in Ukraine and the trauma of visiting the front-lines. My feelings were conflicted with deep admiration for his courage, mixed with serious concern about the damage he had inflicted on us and the wider international fundraising community. By this stage, many of the key figures within AFUO and the broader Australian-Ukrainian community were aware of his actions, openly describing them as fraudulent.

There was a tension in how people now viewed him. On one hand, his reputation had been badly tarnished; on the other, he had genuinely risked his life on the front-lines. Striving to take a balanced perspective, I could see that he had just been unable to hold back, rushing off to Ukraine before we were ready to properly support his ambitions. He seemed to need recognition, portraying himself as a hero on Facebook and social media, with local papers reporting on his exploits. But this came at an unacceptable cost, taking money from others under false pretences.

I had no idea how these issues could be resolved in the long term without further damaging his reputation or undermining the good work he had accomplished in the past. Unable to find an answer, I buried the conflict deep inside myself, hoping the problems would somehow resolve without additional complaints emerging from the public.

I was keen for Lana and Ukrainian Patriots to succeed; they are more than a mere assembly of concerned Ukrainians and international supporters; they constitute a dynamic movement. A dedicated team who work relentlessly to provide assistance and support to those defending Ukraine and the civilians entangled in the conflict. Unlike many other aid agencies, they don't simply dispatch funds; they physically present them on the ground, making weekly deliveries. This enabled them to respond promptly to the most pressing needs, ensuring that donations directly impact those who require assistance the most. Whenever feasible, procuring the supplies locally within Ukraine and Poland to reduce costs and delivery times. In cases where local procurement isn't an option, they expedite supplies from international manufacturers, routing them through a temporary warehouse in Lublin, Poland, before delivering them directly to civilian battalions and humanitarian aid workers inside Ukraine. Out of all of the aid agencies that I had met, like UK Aid, Ukraine Patriots were a group of individuals who were affecting a tangible, real difference in the lives of those most damaged by this ongoing war. More importantly, they were Ukrainian.

Regardless of what Damo had done, like UK Aid, Ukraine Patriots seemed to be one of the best options available to get the majority of aid delivered to those who needed it without losing high percentages of the aid. From a local point of view, they were unique.

At this point, I found myself facing two serious predicaments: the unresolved incident in Lviv involving Smart Medicinal Aid and the fact that Damo had taken all of the company's funds. I was struck by how quickly good intentions could turn sour and felt at a loss as to how these problems could be resolved in the long term.

Meanwhile, Damo threw himself into work with Lana and Ukraine Patriots, reconnecting with many of the volunteers he had worked

alongside in late 2022 during operations in the Donbas region.

The Investigations

In the meantime, I still had not been able to understand why I had been assaulted in Lviv and wanted to understand what I had seen that caused the warning. From January to March 2023, Finn and I began to research Smart Medicinal Aid as an organisation together with their donors during my recovery period. Finn had a lot of local contacts within the Lviv community and was able to gather critical more relevant information and intelligence.

The operational structure of Smart Medicinal Aid was unusual as there seemed to be only one senior "clinician", Irena herself; the remaining people in the office were all very young and unsure of themselves. After a day in the offices with the team, it became clear that she micro-managed the entire organisation and everything that happened daily. During the investigation conducted in late March 2023, we identified a social media post from April 2022 in which Irena claimed that Smart Medicinal Aid had saved 58,000 lives by distributing thousands of tourniquets donated from the United States.

A review of the photographic evidence associated with the post revealed that I had personally observed the movement of hundreds of these boxes on the day of my visit. The Smart Medicinal Aid team had been transporting them behind the ambulances, the same vehicles depicted in my photographs.

This discovery provided a plausible explanation for Bruce's unusually forceful reaction during my visit. It appeared he had not anticipated my presence coinciding with the movement of these potentially illicit items, highlighting significant concerns regarding the legality and ethical compliance of Smart Medicinal Aid's operations at that time.

At first, I was reluctant to fully accept these findings without additional confirmation. However, Finn had far more information than I had anticipated, connecting me to two other firsthand witnesses, including one who had been present the day before my visit.

After speaking with the woman from Ireland, she confirmed that she had indeed been at Smart Medicinal Aid the previous day, inspecting ambulances donated by the Lifeline Ambulance Service and the Irish Medical Mission. Caitlin, who had been there on a separate mission, had discovered a discrepancy with the tourniquets. She had stumbled upon a hidden room with a false wall, which had triggered a full confrontation with Irena. In hindsight, it was clear why Irena and her team had reacted so poorly to my visit the following day, they must have assumed I was there to investigate the dispute from the day before.

I spoke with Caitlin at length, hoping to resolve the matter without further confrontation, but I could not reconcile the issues. Caitlin had even taken photographs of the hidden room as evidence and confirmed with the American suppliers that the tourniquet supplies had arrived from the U.S. in early 2022 and that the serial numbers on the boxes matched the initial delivery, intended for the Ukrainian Defence Force before April of that year.

Having visited Smart Medicinal Aid and the warehouse in November 2022, I wondered why the tourniquet boxes were still there. This discrepancy was puzzling, Irena had publicly stated on social media that they had already been delivered, saving lives. This inconsistency was particularly notable as it was seven months after her initial delivery declaration.

We now understood that I had inadvertently stumbled in to the underhanded movement of the tourniquets a day after the other formal witness from Lifeline Ambulance Service had also visited Smart Medicinal Aid and discovered that thousands of the tourniquets that they had been donated were still at the warehouse hidden in the underground room.

This had triggered the process of them moving the tourniquets to another location, effectively hiding the evidence, at the very moment

I arrived, unknowingly stumbling into the unfolding drama.

As if this wasn't bad enough, more witnesses and former employees of Smart Medicinal Aid had begun to come forward, confessing that they had been bribed, bullied, and intimidated. From an organised crime perspective, the company was an ideal front: posing as a legitimate medical NGO in a war zone gave them the perfect cover to receive large quantities of donated aid, skimming off twenty percent, or more of the profits. What had happened to me was just the beginning. The scale of the truth was going to be even more confronting.

Convert picture taken at Smart Medicinal Aid of the hidden room containing the missing tourniquets, taking a day before my visit. November 2022

Smart Medicinal Aid did not want me to be able to verify that I had seen them moving the medical supplies, something, that should not have been there. Even though I was safe at home in Australia now, I found myself in a situation where I was out of my depth, so I called the Australian government for advice.

As you can imagine, they were very accommodating. They advised against travelling to Ukraine or assisting in the aid sector, suggesting instead that I refrain from future trips, follow-ups with Smart Medicinal Aid, or even continued support of the established aid agencies that we had already begun to support.

Their reaction and advice struck me as disconnected from the Australian spirit of helping others and supporting humanity within the free society we enjoy. I was deeply disappointed by how politically aligned, disjointed, and biased their approach seemed. Australia appeared out of touch, insulated in its own comfort bubble, seemingly unaware of the international pressures driving inflation on supermarket shelves and at the petrol pump.

Understandably, the arguments between Damo, the witnesses me had also caused upset within the AKU, and serious disagreements had broken out within the Australian Ukrainian Federation, (AFUO), whose members had donated significant sums to our organisation. While I no longer felt able to continue supporting the company, another volunteer, a retired Army officer with Ukrainian family ties was willing to assume my position as a director. I was grateful for his support and arranged to meet him to discuss the handover.

I felt compelled to be honest about Damo's unethical behaviour and its impact on fundraising efforts in the community, but I was also overwhelmed with a desire to offload the burden I had been carrying. By the end of our meeting, he was fully briefed and prepared to take over the responsibilities.

Mike appeared highly professional and eager to propel the organisation forward. With deep personal family connections to Ukraine, he proved to be a capable and personable individual, well-suited to assume control. I candidly discussed the ongoing issues with him, but he remained enthusiastic about moving the organisation forward. I provided Mike with a comprehensive copy of the company's founding constitution and all relevant paperwork as I prepared to set my exit date.

I spent much of the early months of 2023 finalising my commitments to Ukraine and trying to plan a new beginning for myself. Yet, unbeknown to me at the time, my experiences in Ukraine had left a deeper impression than I realised and was beginning to evolve into a larger, ongoing commitment.

By now, Damo had returned from Ukraine considerably earlier than expected, staying only six weeks instead of the planned ninety days.

Whilst I wasn't initially concerned, I began to get reports through the network that he had returned earlier due to concerns about my resignation and the misuse of the donated funds. There was nothing I could do to help him as I was already managing all of the complaints that had come through from his actions; I tried to put it towards the back of my mind and carry on.

Then, unexpectedly, a couple of young men who had been working with Damo, Ned, whom I already knew and Tomas, an associate from the AFUO, reached out to me. They began to ask detailed questions about the early days of the NGO Funds for Community Aid, including how the company was created, the funding model we had used, and any background information I had about Damo. At first, I was puzzled by their inquiries and the concerns they raised.

Nevertheless, I had nothing to hide and had always promised full transparency. To honour that commitment, I published most of the company's founding documents on a shared drive for them to review. Still, I was unsettled to learn that several people had already reported the AKU and Damo to the authorities. Even so, I remained steadfast in my decision: if transparency and openness were essential, then I would continue to provide them with full access to all company records and documentation on which the organisation and AKU, NGO had been founded. The only information I withheld was personal and confidential material relating to Damo.

After two weeks of getting to know the two, I received overwhelming news that Damo had over-reacted angrily to my resignation and had devised a plan to prevent us from reporting him to the authorities. He remained unconvinced that there were any external complaints and squarely placed the blame for the tarnished reputation of the AKU and company, Funds for Community Aid, on me.

Ned called me one morning to discuss the challenges he had faced in Ukraine while working with Damo and Ukraine Patriots. Although what he shared aligned with my own experiences in Australia, I was hesitant to escalate matters further by taking any immediate action.

That changed when Ned began sending me copies of emails and text messages from Damo that outlined a plan to physically harm me and intimidate my family. The evidence was indisputable. The seriousness of the situation sank in quickly, and the gravity of Damo's intentions deeply unsettled me. Despite my concern, I was extremely grateful that Ned's conscience had led him to report the plan to the authorities rather than carrying it out.

I now had a verifiable third party confirming what had happened, Damo's aggressive behaviour and plan that had already been reported to the authorities in Australia by complainants. I also disassociate myself from the AKU and his actions. The situation felt like it was spiralling out of control.

Despite reporting all the problems with the AKU and the events to the authorities, most people I spoke to found it difficult to comprehend and were at a loss. The police in Western Australia, either didn't believe me or would have preferred not to be informed in the first place.

During a visit to the police station to follow up on Ned's report, the officers simply looked blank and told us to contact the police in New South Wales, where Damo lived. When we rang the New South Wales police, they seemed even more confused and referred the matter back to the West Australian authorities. It felt like we were being bounced around in a bureaucratic circus, with no one willing to take ownership. The officers appeared unsure of their responsibilities, preferring to misunderstand the situation or wait for the Federal authorities to take the lead.

Despite multiple crimes being reported, supported by specific evidence and witness statements, neither the state police nor federal services seemed capable of understanding, let alone acting on, the allegations.

At the time, I didn't fully grasp how serious this would become. But this failure to act was the start of a much larger problem, one that would come back to haunt me and demand my attention in the future.

I had come to know Ned and Tomas, with Ned ringing me most days, attempting to understand how the NGO had developed through the Australian Ukrainian Embassy seeking support from Australia. Ned was a young, energetic defence veteran and professional soldier who understood military structure, operational tactics and deployment. After serving with the Australian Defence Force, he had acquired additional experience in Europe before moving on to volunteering with the AKU and Australia in Ukraine as a civilian before finally connecting formally with the Ukrainian Legion.

He was socially responsible and committed to integrity, a quality that had caused him problems in the past but, thankfully for me, gave a clear and honest account of the events that had unfolded in Ukraine. Telling the truth could be challenging, especially when dealing with people who would rather avoid the facts. Like Ned, I shared his perspective, was comfortable with his approach, and understood the difficulties we both faced.

In many ways, we had laid the groundwork for a new network of Australian volunteers in Ukraine. Ned eventually returned to Kyiv, serving with the Ukrainian Legion and checking in with me regularly. Meanwhile, I continued supporting established aid agencies and Max in Kyiv wherever I could. Ned focused on developing a professional group of Australians within the region, working to embed drone capabilities as part of their operations.

I also took time to build a relationship with Tomas, the second witness that I had encountered in Australia and member of the AFUO. Like Ned, he was a young Australian-Ukrainian committed to supporting the fight against Russia. He fitted seamlessly into the team, bringing with him a speciality in machine learning and AI, ideal skills for advancing modern drone technologies. Like Ned, Tomas was ethically driven and determined to ensure that as much aid from Australia as possible reached Ukraine.

For the first time in many months, I felt productive again. Working quietly within a capable, like-minded team was both grounding and satisfying.

In the meantime, behind the scenes, Finn had been building a case against Smart Medicinal Aid. He was urging me to return to Ukraine to support his investigation into the NGO, framing it as my social responsibility to follow up on the incident in Lviv. I could see that he was working hard to build and verify his story, and, to be fair, he had a point. He had invested a considerable amount of time into the project.

To push me toward action, Finn even began sending me details of other witnesses; members of the Irish Medical Mission and former colleagues of Irena who had allegedly been bribed or threatened.

By this stage, my head was spinning. I was juggling two separate but equally serious crises, the allegations surrounding Smart Medicinal Aid in Lviv and the situation with Damo, who had absconded with all of our donated funds.

The truth was, I had no idea which issue to tackle first. What I had learned from trying to address these problems was that both individuals and organisations can sometimes struggle to confront the truth, to the point where acting ethically can almost be frowned upon. Working within the international aid sector was proving to be complex, requiring constant compromise and a delicate balance between the demands of total war and the obligation to remain ethical while meeting the requirements of international law.

By this point, Finn had convinced me to return to Poland and Lviv to follow up on the Smart Medicinal Aid incident he had been investigating. He was working with the Security Service of Ukraine (SBU), the country's primary intelligence, law enforcement, and security agency, as well as Interpol. Finn was increasingly persuasive, making it clear that something was seriously wrong with the NGO and that he had built a strong case against them.

He planned to introduce me to several others who had also worked within the organisation who possessed convincing evidence of aid being sold under the counter, which they had already submitted to the Ukrainian authorities.

For me, it was simple, I just wanted to understand what had happened and why. Once I had saved enough funds for the trip, I booked flights back to Warsaw to meet them in the spring of 2023.

Australian Police

As I began to assemble my equipment in preparation to return to Ukraine, I began to question the sanity of returning to the war zone, feeling that I had only just survived the first trip. However, both Finn and Max were delighted with the news and had already started to plan activities in anticipation of my visit. Max had started to arrange a number of aid trips from Kyiv to Bakhmut in the Donetsk Oblast to alleviate a severe need for supplies, whilst Finn had planned a trip back to Lviv to finally tackle the issue of the missing aid that had continued to disappear under the care of Irena and the Smart Medicinal aid team.

Early one morning, a week before I planned to leave for Poland, the front door rang and everything changed for me instantly.

Stunned by what I was faced with, all I could do was remain polite to the police who stood at my front door, waving a search warrant, and invite them in. The senior detective in charge then explained that I was being arrested because they believed Ned, Tomas, and I had fabricated the narrative surrounding the allegations against Damo and the evidence we had submitted.

Though I was shaken by their accusations, I calmly made a cup of tea while they gathered and impounded all of my electronic equipment and correspondence, before taking me to the police station for an interview.

By now, I was seriously swamped by the day's developments, yet I remained confident that the truth would prevail. I was prepared to share every detail of the events in Ukraine and Damo's misconduct, particularly how he had misused all of our company funds and fraudulently taken more from the public. Despite spending hours explaining the intricacies of the past year, it was clear their minds had already been made up. I was charged with "creating a false belief", in other words, being accused of lying.

At that point, I denied the accusations. The senior sergeant reacted sharply, shifting in his seat as he slammed the papers onto the desk and raised his voice. I held his gaze calmly, noticing the anger and frustration in his eyes. He seemed intent on forcing a confession out of me then and there, but his efforts had failed.

Shortly after being returned to my cell, the gravity of the situation began to sink in. I didn't understand why they distrusted the evidence from Ned and Tomas, yet they appeared intent on securing an intimidating conviction. They covertly pressured me, insisting that I was "now in their world and had to accept it."

The whole incident felt utterly bizarre, almost like being back in the trenches in Ukraine. Why would a middle-aged chaplain, freshly returned from providing aid in Ukraine, fabricate lies about money stolen from an NGO? And why would others outside the organisation submit evidence to support those claims unless they had substance?

The allegations against me were ridiculous. I could already see that the senior sergeant had grown unsettled by my calm denial, seemingly invited to prove it. All he had to do was examine the evidence and trace where the money had actually gone, from public donations, into Damo's personal account.

Most importantly, I had no motive to make factitious allegations against Damo. Ned had been the first to report him, followed by Tomas, and I had been simply obliged to support their accounts with my own.

This was just another day during a difficult period of my life, something I would have to face and move forward from. It quickly became clear that (WAPOL), the West Australian Police, had been unable to properly process the allegations and the evidence provided. They appeared to focus inwardly, rather than considering the facts in a broader context, including information from the Embassy and other sources.

My priority was to regain my passports, ready for my return to Poland as I now only had days to go before I was expected back in

Ukraine and needed to concentrate on the commitments that I had already made.

The judge in court the following day, seemed both confused and compassionate with my plea to have my documents returned as I had already booked return flights from Poland well within a ninety day time frame and had explained that the visit was a follow up journey in support of an international NGO working through the Embassy. Whilst the police prosecutor insisted that I could be a flight risk, I could see that the judge remained confused by the severity of the bail conditions in relation to the scale of the charges that were menial.

All of my documents were returned and I was released with a pocket money bail. The frustration in the prosecutors eyes was obvious and at the same time, I just couldn't understand the scope or details of the evidence against me, it just didn't make sense.

From this point, I needed to focus on my priorities; to return to Ukraine, honouring my commitment to my colleagues, the friendships that had formed, and the established NGOs.

Their need was greater than mine, regardless of my current circumstances.

CHAPTER 11

2023 Back to Ukraine

With enough money to return, I remained focused on the plans we had already made travelling back to Warsaw from Australia, reluctantly repeating what I had only recently done only months before, knowing that I could only stay for a short period of time and complete a limited amount of work. The situation with Damo had soured relationships with Ukraine Patriot who were more concerned about covering up what had happened rather than tackling the issue at hand. Whilst I could understand how they felt, I couldn't be associated with the AKU or Ukraine Patriot until they had come to terms with what Damo had done and the inappropriate use of donated funds. War is complicated enough as it is fighting the Russians; we didn't need to be arguing amongst ourselves and stealing money from each other. From, another point of view, the war itself had created an atmosphere of desperation which had led to distrust.

Once again, the flights from Australia to Poland proved long, exhausting, and physically demanding, spanning vast distances and multiple time zones. The journey required me to brace for yet another arduous travel marathon, a prospect that, at this point, felt increasingly unappealing and made me question the sanity of my actions. I was now taking an unacceptable level of risk. I didn't fully understand why. I felt angry, frustrated, and numb from the recent developments in Australia, yet compelled to complete the mission I had begun the previous year: to address the problems that had arisen and to understand what had gone wrong at every stage of the journey.

The initial leg of the journey, from Australia to Qatar, proved to be the most challenging. The flight felt even longer than the first time. By the time I reached Qatar, serious doubts had crept in. While I understood that Interpol and Finn required evidence regarding my experience at Smart Medicinal Aid, I couldn't help but be gravely concerned about the threats I had faced at home and the risks I was now exposing myself to. Finn had assured me of our safety and had engaged a professional Ukrainian-speaking "fixer", which did boost my confidence. Nevertheless, the full story would only emerge when Finn eventually released his documentary, my participation was only a key part of his story. Experiencing a strong sense of déjà vu, I went through the familiar airport and customs routines.

When I finally arrived, I was simply relieved to be there, ready to complete what we had set out to achieve.

As soon as we landed in Poland, I found myself constantly scanning my surroundings, relying on my original trade-craft training from my Army days decades earlier. I checked for anyone following me, pausing to glance into shop windows to reassure myself that it was safe.

I had already booked into a central hotel in Warsaw under two names to make it harder for anyone to track me. I stayed in the second room under a pseudonym, barricading the door for added security.

After a couple of days, the team assembled. Finn flew out from Ireland while Andrew from the Irish Medical Mission, the initial witness that had started the Smart Medicinal Aid investigation, travelled from Ireland in a modified van that could be used as an ambulance. His journey had started long before mine as he was driving thousands of miles across Europe, land and sea.

It was reassuring to see Finn again; his enthusiasm was infectious, even though I knew his drive to get a story could sometimes be reckless. Sitting at the bar, sharing a beer, helped calm my nerves listening to his soft Irish accent, though I instinctively chose a booth at the back of the room, facing all the entrances with a clear escape route planned.

Andrew arrived much later, understandably tired after such a long drive. To make him comfortable, I had booked another room and laid on dinner, accompanied by the obligatory pint of ale. Although it was late, we needed a couple of hours to connect before driving back into Ukraine the following day. I had to get to know him and trust his judgement before risking my life.

Andrew was quietly confident, a typical southern Irish gentleman. After a couple of hours discussing the plan for the days ahead, along with a brief synopsis of his experience with Smart Medicinal Aid, I trusted his judgement. We seemed to be on the same page, with no hidden corners, even though his account raised more questions than answers. After everything he had seen and reported, how was Irena still getting away with her actions?

Andrew's rapport with Finn was also reassuring. They clearly knew each other well and had worked extensively together before, which added another layer of confidence as we prepared for what lay ahead.

Andrew had begun his work in Ukraine a year earlier with Smart Medicinal Aid, joining the company at its inception. He collaborated closely with Irena, the CEO, eventually ascending to the position of Executive Officer and second-in-command. With a background in building NGO's from the ground up, he demonstrated natural leadership and professionalism.

According to Andrew, his partnership with Irena ran smoothly until he uncovered the blatant misappropriation of two ambulances during a transfer convoy. He presented me with video evidence showing nine ambulances departing, clearly capturing their number plates, but only seven arriving at the destination, two had gone missing. When he inquired, he was told that one had been hijacked and the other lost. In reality, they had allegedly been sold on the black market for personal profit.

When Andrew confronted her, refusing to participate in the corruption, she instantly dismissed him, tarnishing his reputation. Although Irena had offered him wealth and a secure future, his consciences could not tolerate the corruption and chose to leave. Unfortunately, no one was willing to listen to his concerns. In many ways, by taking the most ethical path, it had almost destroyed him. My situation was much the same as his; I had faced the same dilemma. By reporting Damo, I became a victim of isolation, threats, and even prosecution by the local police in Australia.

I greatly respected Andrew's honesty and dedication to the truth, but I began to perceive a troubling reality: those in authority often preferred to remain ignorant, as acknowledging the truth risked exposing their own indifference, wrongdoing, or complicity.

Andrew had now returned to Ukraine to meet up with us and his Ukrainian wife, taking us back to Lviv on the way and to begin a new medical evacuation service for casualties, moving them from the orange area of a combat zone to hospitals in safer neighbourhoods.

The more I got to know him over a couple of days, the more I liked him. He had a genuine passion for his work and was a natural life saver. He wasn't interested in his own wealth as long as he had enough to make ends meet, feed his family and run the ambulance. Andrew had a network of friends, colleagues and associates all over Ireland and Ukraine helping most of them himself when they were unwell asking for nothing in return. It is rare to meet people like him in life and important to study how they see themselves, not as heroes, but as normal human beings born into a wholehearted way of life providing care for others. He was also a natural leader and well organised, able to confidently cope with most situations. I liked him, he exuded the best of what humanity could offer.

Finn briefed us on the developments since our last visit. A local undercover operation, carried out off the books by a Ukrainian battalion commander, had uncovered limited evidence but pointed to a troubling reality: the Smart Medicinal Aid team in Lviv seemed to be linked to an organised crime group (OCG) that had allegedly cultivated ties with the Australian-Ukrainian Embassy, using embassy officials as cover, whether knowingly or not. Finn had already sent me intimate photos of Irena with a senior embassy official, and this only heightened my concerns. It added another unsettling layer to the situation, showing how deeply she could influence individuals at the highest levels of government. Irena was a master at playing the game, using her skills, charm, and allure as an attractive woman to further her own agenda.

The news was unsettling. It became clear that we could only trust the people we already knew, the battalion commander and Finn's new fixer, Garik.

After a day in Warsaw, we planned the next couple of weeks and decided to meet the investigative team in Lviv within a few days. Early the next morning, we drove from Warsaw down the S17 to Lublin, taking a break before continuing to Przemyśl, where we stayed at the Restauracja SPA Hotel Gloria on the border.

The location was one of the main crossing points into Ukraine, yet the hotel was astonishing, comfortable rooms, a large spa, and a building team completing the roof. A new supermarket and spa hotel had replaced the refugee tents and aid agency soup kitchens, a stark reminder of how life went on amid the conflict. The scene highlighted the local community's acceptance that the war would persist for years, contrasting sharply with the ongoing human suffering just across the border.

Businesses in Poland had seized the opportunity to profit from the war, tapping into the steady flow of travellers and refugees crossing the border. Accommodation was offered in three tiers at very modest prices: a basic room with hot water for around $10 USD; a step up that included a full breakfast for $15 USD; and, at the top end, a suite of rooms with a lounge for $22 USD, with the option of a spa dip and sauna for just a couple of dollars more. Staying in the suite felt somewhat decadent, yet it offered the best value for money when shared between the three of us.

For those who couldn't afford a room, the choices were bleak, crawling into a cardboard box behind the supermarket and surviving on a cup of soup and a sandwich bought with a handful of coins. I wish I could say I felt comfortable in the relative luxury of the hotel, but it was hard to ignore the sight of elderly women shuffling past with overloaded shopping bags, dragging weary children behind them. Still, I tried not to dwell on it; we couldn't help everyone.

In contrast, we spent a pleasant spring evening in the nearly empty hotel, sharing a few beers as we spoke openly about the issues at hand. Andrew delved into his background, recounting his experiences at Smart Medicinal Aid. He showed me pictures and videos of missing aid supplies, along with images of staff members who had abandoned the organisation, fearful of the repercussions. In turn, I shared my own experiences, describing the injuries I had sustained, showing him photographs of my broken arm and the handcuff marks.

It was clear that the NGO had serious internal issues, yet nothing had been resolved. Smart Medicinal Aid operated with a tough veneer, seemingly untouchable. We resolved to do whatever we could to connect with the local officials who had begun investigating the company and registering witness statements, at least to try to curb the corruption and support Ukraine.

We also learned that some of the organisation's existing suppliers were starting to uncover misappropriated aid themselves reporting it to the UK and Irish-based donors. The outlook for Irena was slowly shifting as more people began to catch on to their underhanded operations. Even if Ukrainian authorities were unable to stop them, we could at least dispense our own form of justice by exposing their greed and the misappropriation of aid.

The next day, we got up as early as possible; Finn wasn't an early bird and tended to resist our encouragement to get up for breakfast until he had a strong coffee with a cigarette. We crossed the border as early as possible to avoid traffic. While we had been hopeful that the crossing would only take a couple of hours, the border guards made us wait in line with some commercial traffic instead of prioritising the ambulance we were driving. Either way, we got through well before lunchtime and moved steadily to the Sokilnyky district in Lviv.

Sitting in the van at the border, I almost triggered an international incident when I leapt out of the vehicle and dashed towards the restroom in desperate need. As I sprinted to the Porto-Loo, I could hear a guard shouting at me, though her words were incomprehensible. Hoping for the best, I rushed into the toilet and closed the door. Fortunately, I wasn't met with gunfire through the door, and I could hear Finn and Andrew trying not to laugh explaining my urgent bathroom situation to the guards. Thankfully, by the time I emerged, the situation had calmed down enough for me to pull up my pants, as they accompanied me back to the vehicle, wearing amused expressions on their faces. Finn and Andrew found the entire incident understandably hilarious.

As soon as we crossed the border, the awareness of being back in Ukraine hit me hard, and I needed a couple of hours to readjust. I didn't want to be back, but wanted to finish what I had started. More importantly, by now, I had been supporting Ukraine for nine months and had developed a deep relationship with its people and culture. The country deserved my commitment.

The mood in Lviv was calm, with shopping malls bustling with everyday life as people moved through. We drove along the ring road to the southwest of the city, heading for our agreed meeting point. There, in the food court of a central shopping mall, we sat down for lunch. Although I felt relatively at ease, I couldn't help glancing over my shoulder, alert, looking for anything out of the ordinary.

Back in Lviv

We said goodbye to Andrew as it was his time to move on and start work and waited for Garik, our local fixer. I was sad to see Andrew go as I felt comfortable in his company; he had given me considerable hope and insight into Smart Medicinal Aid and how the aid sector worked in Ukraine. Whilst it had unsettled me that the company and officials had mistreated him, he was highly resilient. He had given me hope that many other people knew the truth and that I was not alone in what had happened to me.

Garik arrived shortly after. He was a seasoned Ukrainian fixer with combat experience and a background in journalism, fluent in English, Ukrainian, Polish, and Russian. An ideal guide and translator, he carried himself with quiet confidence. Unlike Finn, Garik was impeccably dressed and composed, every inch the professional. His presence projected toughness and competence, which reassured me and gave me confidence in his abilities.

The relaxed atmosphere in Lviv contrasted sharply with the crowded scene inside the shopping mall, where hundreds of young men of service age mingled freely. This disparity sparked a discussion about the differences between Lviv and the eastern districts, or Oblasts of Ukraine, including Kyiv. While avoiding a definitive conclusion, it was clear that fewer men from the Lviv region had been conscripted.

Garik, our fixer from Odessa, offered a more informed perspective. He explained that the issue was largely political. Families from Lviv Oblast, with their relative affluence and stronger political connections compared to those in less wealthy regions, faced a lower likelihood of conscription into front-line combat roles. Protected by influence and geography, they lived in a region shielded from some of the harsher realities faced elsewhere in Ukraine.

A distinct separation existed between different areas of Ukraine in terms of conscription rates, with Lviv appearing partially immune. Emphasising this contrast was a wrecked four-by-four vehicle in the shopping mall's entrance hall, riddled with high-calibre holes, suggesting the obliteration of both the car and its occupants.

Adjacent to it sat a fundraising bucket, a sight I had only encountered in this location. It almost seemed as if Lviv was disconnected from the main war, soliciting donations to support the war effort.

The Battalion Commander

After leaving the shopping mall, we travelled further south of Lviv to a discreet military site where Finn introduced us to his local contacts and a battalion commander. Many of them were Ukrainian military personnel and drone developers.

I was excited to meet new contacts, including several who crossed the border daily from Poland. Our discussions centred on the development and deployment of drones in the war, the pressing needs of aid agencies, and the most practical ways to support Ukraine in the months ahead. As always, we used the opportunity to connect Australian aid agencies and developers with local partners. We remained with the team at an undisclosed location in the rural outskirts of Lviv.

The battalion commander was an outspoken, effervescent character with a generous nature carrying a number of facial scares from earlier encounters with the Russians. He hosted us for a sumptuous dinner, complete with copious amounts of alcohol, which created a relaxed, convivial atmosphere and made it easier to tackle difficult conversations. Over the course of the evening, we listened to countless war stories, accounts of how many Russians the team had eliminated and the hardships they had endured.

Many of them had been fighting since 2014, from the time of the Crimea annexation, and were now focused on drone development and ensuring aid was not misappropriated. They believed mismanagement only complicated their work and jeopardised the steady supply of the parts they needed for their drone development.

They emphasised that the international community needed to see them actively engaged, not only in the fight against Russian forces but also in countering the rise of organised crime, an issue they had to navigate carefully, as such groups often carried powerful connections. As a team, we also received an update on the situation with Smart Medicinal Aid and were cautioned to remain aware of its influence and political leverage, particularly at a local level.

It was a productive couple of days, during which I gained valuable insights into the Ukrainians' most pressing needs, more importantly, witnessed just how much their technology had advanced in under nine months. The progress in drone development, AI integration, machine learning, and mobile command control vehicles was remarkable. I became acutely aware of how crucial it was for Australian, UK, and American suppliers to connect with these drone developers if they were to provide equipment that was truly effective and beneficial.

That evening, Garik served as our interpreter, since neither Finn nor I spoke Ukrainian, and the Ukrainians could understand little of what we said, aside from a couple of younger team members who had learned some basic English growing up. Though the language barrier slowed the exchange of information and ideas, the atmosphere remained relaxed as we ate and steadily working through bottle after bottle of wine.

Most of the conversation was easygoing until the topic turned to Smart Medicinal Aid. At that point, the commander and his men grew louder, more animated, and the room quickly filled with tension. At first, I couldn't tell whether the change in mood came from the alcohol or the subject matter. Fortunately, Garik's professionalism shone through; he kept the conversation steady, even as the commander's agitation rose when Irena and Smart Medicinal Aid's name came up, along with accusations of them stealing aid from the community.

As the details unfolded, it became clear that we were dealing with a far bigger issue than we had initially anticipated. Irena had entrenched herself deeply within the "Catholic Orthodox" community in Lviv, which not only shielded her but also provided a certain level of impunity for her actions. In the chaos of wartime, it seemed almost permissible for her to skim from the aid supplies, so long as the Ukrainians received eighty percent, the missing twenty could be brushed off as "operational costs." This meant diverting thousands of tourniquets to the black market or disposing of donated ambulances and high-value medical equipment to criminals.

As a Chaplain, I found her association with the Catholic College utterly repugnant. The college's walls offered her an unquestionable level of immunity, whether or not they were aware of her actions. With the growing body of evidence against her, it was only a matter of time before the truth would spill out, exposing the college to serious criticism and tarnishing its reputation.

When the commander finally reached the limit of his frustration, Garik deftly shifted the subject back to something guaranteed to lift the mood, killing Russians. He pulled out videos of himself and his team trench-clearing in the Donetsk region: moving in pairs through bunkers, methodically cutting down conscripts as they scrambled from corners or bolted like frightened rabbits, all caught on body cameras. The footage electrified the room and, for me, added another layer to my perception of Garik. Beneath his icy exterior was a man hardened by skill and experience.

For all my unease, I couldn't help but feel reassured by his display. Thinking of my own survival and the uncertainty of the days ahead, I asked Garik for his contact details and information on the group he worked for, hoping I might one day be able to call on their support, should I need it.

Police in Lviv

After several days working alongside the battalion commander in Lviv, we resolved to complete the task that had first brought us there. We gathered our documentation and formally submitted the evidence concerning Smart Medicinal Aid to the local police authorities. It was not an act driven by confidence in the system; experience had already tempered expectations. Rather, it was a matter of principle. We knew the outcome was uncertain and that the machinery of wartime bureaucracy moved slowly, if at all. Yet there remained a quiet, stubborn hope that someone, somewhere within the process, would take the time to listen. Finishing what we had started felt less about immediate justice and more about drawing a line, about being able to say that we had done what conscience required, even if the wider circumstances suggested it might change very little

We began the morning at the central Lviv police station, speaking with a young officer at reception. Garik quietly explained our purpose and offered to show him our evidence folder, which contained statements and photographs. At first, the officer seemed interested, but as the seriousness of the situation and the details of the report became clear, he grew visibly uneasy. He disappeared several times, returning each time to ask for more information, before ultimately referring us to a more appropriate local station in the correct district near the aid agency's campus on the Catholic

College grounds.

We finally arrived at the correct police station, Sykhiv District Police Precinct, late in the afternoon, after visiting two others earlier in the day. By this time, concerns were growing that we might have run out of time. At reception, a young woman summoned a senior officer to take our statement.

Tensions quickly escalated as we presented the evidence to the senior officers. Garik, however, remained composed, speaking softly and steadily. As the afternoon wore on, more senior officers joined the discussion, each offering various reasons why we could not proceed with reporting the incident. It became increasingly clear that they were unwilling to acknowledge the incident or the evidence we had brought.

When our discussion concluded, we faced the district commander, who declared that it was too late to report the crime. Throughout the room, each officer had offered a different reason why our report was no longer relevant. Their words and actions made it clear they were reluctant to take any action. Despite our repeated requests, none were willing to provide their names or accept any documents. They were clearly familiar with Smart Medicinal Aid and more concerned with avoiding involvement. To be honest, I found myself beginning to understand their perspective, the risk of being associated with the case or confronting Irena locally felt far too great.

As we left the station, we managed to convince a middle-aged senior officer to accept a copy of the statement for reference, fully aware that no action would be taken. It was likely that the report would simply inform Smart Medicinal Aid that we had visited.

Once back in the car, the situation became painfully clear: the influence of Smart Medicinal Aid, and Irena, was so extensive that it even unsettled the police. There was no possibility that any report of misappropriated aid would ever result in action.

From that moment on, we faced a significant risk. Smart Medicinal Aid would inevitably discover our presence in the area and respond within hours. Before I could fully process the situation, I felt as though I had been thrown back to the Cold War, hunted and tracked by the KGB. I glanced at the others, searching for a clue to their thoughts, amazed by their composure.

Garik had accepted the circumstances openly, fully aware of how far things had deteriorated. The authorities would never take our evidence seriously. To him, this was routine, a demonstration of his consummate professionalism; calm, collected, and precise. As a native Ukrainian, he understood the stakes instinctively. We needed to move swiftly, leaving Irena's sphere of influence and the Catholic College behind. She had strategically embedded her operation within the college, entwining it with Smart Medicinal Aid, using the institution both to shield her position and to strengthen her leverage within the local authorities.

On the other hand, I could feel Finn's frustration with the outcome, particularly after he had convinced me to travel so far to follow up on the evidence and the statements of others. Finn, Garik, and I had returned to the car and got away, ensuring that we had not been followed, looking over our shoulders from time to time and walking around the block, including an alternative route back to the car.

We drove away at speed, deliberately taking several detours and doubling back to ensure we weren't being followed, only allowing ourselves to relax once we were confident we were safe.

The mood had shifted to disappointment. We all felt disheartened by the outcome, though none of us were surprised. For security, we agreed to part ways. Finn would push deeper into Ukraine, heading toward the front-lines to gather the last of the footage for his documentary, while I had only one priority, leaving Ukraine quickly and safely. Remaining in Lviv was no longer an option. I had to stay on the move, vanishing once more into the distance, beyond anyone's reach.

We drove out of the city in silence, the weight of failure heavy in the car. As night began to fall, we pulled into a service station on the outskirts of Lviv, the dim lights casting long shadows across the forecourt. There, under the cover of darkness, we set about formulating our next move.

As part of the deal for my return to Ukraine, Finn had promised to protect me and get me out if his plan went sideways. What he didn't know was that, while I was eager to cross the border, I'd also been in contact with Max in Kyiv. He was on his way to meet me in Kraków to assemble a large aid run to Bakhmut. Before that could happen, though, I had to be seen disappearing across the border, only then could I return under a different passport and newer identification.

I had now descended into the murky world of undercover work as a dual citizen, swapping phone SIM's and changing IDs to shore up a fragile anonymity.

Evacuation

Finn reached out to several of his contacts in the area and quickly located a young German medical team preparing to leave Ukraine that day. They were taking a southern route toward Kraków, a perfect opportunity for me to slip out unnoticed, surrounded by people the group in Lviv would never think to track. It felt almost too perfect, as if Finn had orchestrated it all in advance.

We rejoined the ring road and pushed west to rendezvous with the team. By now, dusk was settling over the city, and hunger gnawed at us; none of us had eaten since morning. The promise of food, however modest, was a welcome pause before the next move.

Frieda and Hans were younger than I had expected for people who had spent so much time working at the front. They arrived in a battered Dacia Duster, converted into a medical vehicle, its bodywork caked in mud and the obligatory subdued red cross painted across the doors. The car was packed to the rafters with luggage, every inch of space claimed.

It was immediately clear that Frieda knew Finn well. The moment she saw him, she leapt up with a wide smile, calling out, "Irish!" in a German accent before embracing him warmly. We settled into the truck stop restaurant, where steaming bowls of traditional

Ukrainian "borscht" were brought to the table. The sour beetroot soup, with a deep red colour so distinctive, had been one of my first meals in Lviv on my initial visit. Ukrainian food was always vivid and memorable, made from fresh, locally grown ingredients, and offered at excellent value.

My personal favourite was "varenyky", soft dumplings filled with meat, potatoes, mushrooms, vegetables, fruit, cheese, or berries. They reminded me of the "dim-sums" I had grown fond of back in Australia, but here they carried a richness and heartiness that spoke of Ukraine itself.

For the first time in days, I allowed myself to exhale. Eating local food among friends felt like a fleeting sanctuary, a fragile relief before the road pulled us back into uncertainty. Frieda and Hans were a delightful young German couple. Frieda, a newly qualified nurse, had spent more than a year working at the front and was finally returning home for some much-needed rest. We spoke for a while, explaining my situation. They listened carefully, fully grasping the risks and the urgency of my need to leave the country.

What struck me most was their willingness. Though we were asking them to put themselves in danger, they didn't hesitate. Instead, they agreed to drive me across the border and leave me in Kraków overnight.

They understood that the faster I disappeared, the safer it would be for everyone.

We repacked the car with my kit, making enough space for me to squeeze into the back seat. The drive from the truck stop to the border would give us time to talk and settle into each other's company. My German was limited, but Frieda and Hans's English was, as I expected, excellent. By the time we reached the border a couple of hours later, we would have shared more than enough conversation for me to feel at ease with them. My attempts at basic German drew frequent smiles and laughter, a welcome lightness amid the tension of our journey.

We left as quickly as we could after eating, saying goodbye to Finn and Garik with heavy hearts, urging them to stay on the move and keep out of sight in Lviv. As we pulled away, I couldn't shake the worry for their safety, yet I took some comfort in knowing that Finn's boundless enthusiasm, paired with Garik's quiet tenacity, might just see them safely through the months ahead.

The drive to the border was deceptively pleasant. A soft spring evening lightened the mood, and the miles slipped by faster than I expected. But the moment we reached the crossing, everything changed again.

The ordeal was awful. Frieda, though registered as an aid worker, had overlooked one crucial detail, her car had been in Ukraine for more than twelve months without the proper paperwork. That oversight cost us a staggering one-thousand-dollar fine. I was stunned that the Ukrainians would punish a nurse and volunteer who had risked her life to save their soldiers with such a penalty.

Every minute we lingered there increased the danger. If the guards connected me to Smart Medicinal Aid, we would be trapped. I dug into my money belt, pulling out my emergency stash of American dollars. Frieda and Hans added what they could, and together we scraped together the fine.

The change was immediate. As soon as the money appeared, the guards' mood softened. Hours of tense discussions and paperwork finally gave way to the moment we needed, crossing the border, shaken but safe, grateful to have slipped through in one piece.

The tension of the past few hours dissolved as we got back into the car and began the drive toward Poland. I could feel the shift in our situation, and in Frieda's attitude as she gave me a gentle hug in gratitude. We exchanged looks, reassured by the conviction that we had overcome a serious problem that could have kept us detained at the border for days.

I felt a profound sense of relief having been able to help Frieda. She was a shining example of how war can bring out the very best in people, embodying the finest qualities of humanity. In her mid-to-late twenties, she had served as a qualified nurse, dedicating an entire year to volunteering at a front-line first aid station without a single break. Day after day, she confronted unimaginable horrors.

While I had glimpsed some of the darker aspects of war, Frieda had endured them constantly for a full year. The long drive from the border to Krakow gave me time to reflect on the events of the past weeks and to become more familiar with Frieda and Hans' story.

Han's, a graphic designer preparing to join the press corps, had accompanied Frieda for several weeks to document both the war and her journey. Though he insisted his time in Ukraine was driven solely by journalistic ambition, I sensed that his admiration for Frieda's humanity and kindness ran deeper than he admitted. Like her, Han was dedicated, focused, and remarkably well-balanced, yet he carried a quiet burden: the Russian annexation of Crimea weighed heavily on him. He believed it had set in motion a chain of destabilisation that threatened not only Ukraine but the world at large.

As we travelled, our conversations often returned to the war and

Russia's posture toward Ukraine and Europe. We shared the same conviction, that even if Russia were to overwhelm Ukraine militarily, Europe could not, and should not, accept such an outcome. With nations around the globe increasingly fractured and forced to take sides, it was difficult to escape the sense that a wider conflict was not only possible, but perhaps inevitable.

It was a late spring evening, the air warm and carried by a gentle breeze. After so many years in Australia, I had almost forgotten the distinct music of European birdsong, now filling the countryside as we drove with the windows down. For the first time in months, I felt able to exhale and truly breathe. The beauty of the scene brought a rare sense of calm, a reminder that I was finally on the road to safety, leaving behind the complications that had entangled me and the horrors of war that still lingered in my memory.

After a long drive, we arrived in Krakow by the early morning where I had booked myself in to the Ibis hotel for a couple of days rest before Max arrived.

Frieda had gone well out of her way to bring me to Kraków, driving straight into the city centre and leaving me at the doorstep of my hotel. I was struck by how quickly people could form strong attachments in war, built on a foundation of mutual respect, another strange by-product of conflict. She had been close to tears at the border when I handed her my last stash of American dollars to get us across. She knew that I could have chosen the simpler option of walking over the footbridge and catching a bus or taxi on the other side, but I had explained to her that the moment she had offered to risk her life for mine, we had become bound together as friends forever.

Hans helped unload my bags, and together we shared a final embrace, a group hug heavy with unspoken gratitude and the sadness of parting, before we went our separate ways. I was totally shattered, swamped with fatigue and in desperate need of a rest. I ate what I could and slept for over thirty six hours.

CHAPTER 12

A Red Zone Run

A couple of days later, I was still half-asleep when a knock came at the door. Opening it, I was greeted by Max's wide, beaming smile, just as enthusiastic as ever. We began the morning with a late breakfast, speaking quietly in a corner where no one could overhear us. Over coffee, Max laid out the full details of an urgent aid run to Bakhmut. What had started months earlier as a modest plan, a couple of trucks at most, had grown into a full convoy; several vehicles, our Mercedes ambulance from Kyiv, and a number of recovery vehicles.

The sheer scale of the mission took me by surprise. Delivering aid into a city on the brink of collapse was no longer just risky; it was reckless, bordering on madness. This would be one of my last trips into the city, I told myself. I couldn't really explain it, here I was, safe in Kraków, just across the Polish border, with every reason to go home. Yet, the moment Max laid out the plan, I knew I couldn't walk away. He was more than a teammate; he was like a brother in arms. Saying no to him was never an option. I was going back to the front-line and back in to the "Red Zone" on a "Spicy Run".

The fact that we already had a small Australian medical team on the ground, Ella, a nurse from New Zealand, and Ned, a seasoned Army medic, changed everything. They didn't have their own vehicles, they were effectively stuck, and I couldn't just leave them to handle things on their own. They were vulnerable, and there was no way I could let them fend for themselves. They needed to be evacuated, and I couldn't turn my back on that.

The thing that worried me the most was how I'd explain the whole situation to Michelle, my wife. I'd already told her I was only going back to Ukraine as far as Lviv, never mentioning Bakhmut. I had no clue how she'd react when she found out the truth. But I knew that conversation would have to wait for a couple of weeks.

Max, already rested and eager to get moving, urged us forward. So I packed my bags once more, switched to my UK passport, avoiding the risk of my Australian citizenship being flagged at the border, and climbed back into the ambulance without a second thought.

It seemed ironic that we travelled along the same road that I had just been down only days before, but I felt more comfortable than I had expected with a renewed strength in Max's company, the team was back together.

By the time we neared the border hours later, I had convinced Max to drop me off first, closer to the footbridge a couple of kilometres away, so I could cross separately while he drove the ambulance back to the main road and crossed directly.

At this point in the war, we had both learned from our mistakes and knew how to make the most of every situation. An ambulance approaching a border crossing into Ukraine, full of medical aid, was understandably welcomed and given priority, so we had filled the back of the vehicle with bandages and basic supplies from our contacts in Warsaw and Kraków, having planned ahead.

Meanwhile, I had also double-packed my backpack with critical medicines bought from the same Polish associates and carried the letter of association from UK Aid. Anything that could link me to the Ukrainian-Australian Embassy was hidden away in a concealed, stitched-in pocket, difficult to find.

My only real concern was that one of the guards at the crossing might recognise me. But the footbridge was far enough from the main road that it seemed unlikely I'd run into the same faces.

The plan was solid, and it worked. I stood in the queue for forty minutes before the guards scanned my passport for previous stamps but found nothing, with no record of my earlier crossings. Satisfied, they waved me through without hesitation, I was just another medic slipping across the border. They didn't even bother checking the contents of my bags in depth. If anything, they just looked relieved to see another volunteer.

I had crossed the border well before Max, giving me enough time to stop in the car park and make a brew on the small camping stove I'd started carrying everywhere. It took Max another forty minutes to get through and drive down to pick me up. Although he was willing to pause for a quick drink, I could sense his urgency to keep moving, we still had a long nine-hour drive back to Kyiv and on to our new location near the warehouse.

The quickest route to Kyiv from where we were led us back along the E40, past Lviv and around the ring road. The closer we drew to the city, the more uneasy I became, refusing to leave the ambulance until we had cleared Lviv and moved on toward Rivne. Yet the morning was bright and full of promise, deep green fields stretched out beside the road, broken by small industrial zones and scattered housing. Talking with Max lightened my mood and steeled my resolve to reach Ella and Ned. More importantly, the aid run to Bakhmut had now grown into a massive operation, thousands of tonnes of supplies were destined to keep the city standing a little longer and to rescue as many of the wounded, women, and children as possible. It was going to be a large operation, involving a number of our original colleagues and contacts from the Dream hostel in Kyiv.

We stopped near Rivne at a roadside café for a meal, and afterwards I took the wheel again. It felt good to be back in control, surprised by how quickly the ambulance felt familiar again beneath my hands. With the window down and the fresh air rushing in, I settled into the seat, a growing sense of anticipation carrying me forward into yet another unfolding chapter of this adventure.

The Reality of War

As the day wore on, Max continued to brief me on the situation in Bakhmut, outlining the coordination he had established through battalion commanders across each Oblast. In turn, I told him about the Australian medics who had been working at the front for months. From my conversations with Ned and Ella, I knew all too well the grim reality they faced. Both were exceptionally capable and resilient, but the casualty rate among daily patrols had surged to more than eighty percent, a figure that was almost impossible to comprehend.

It had become clear that Max relied on my support, not only for my embassy connections but also for my experience. Fortunately, our goals aligned perfectly. Bakhmut was now the very heart of the conflict, with casualties rising daily. I spoke with our team every day, offering what encouragement I could, but the truth was stark: even the most basic medical supplies and food were becoming harder to secure, and their situation was growing ever more precarious. We needed to get them out.

To make matters worse, the harshness of the Ukrainian conflict had taken hold, turning movement into a daunting task. Most of the roads were buried under rubble and strewn with road blocks, making travel treacherous. Any trip to the city would require meticulous planning and the right equipment to avoid disaster.

The city held strategic importance for the Russians, granting them greater access to the east toward Kramatorsk and furthering their overarching objectives. As spring had set in, the conflict had moved from an air offensive to a ground assault campaign, relying heavily on artillery and drones. This approach would continue until the season changed again restricting the deployment of heavy armoured units as the ground either became too muddy or frozen solid with a permafrost. The conditions were brutal, a scorched earth littered with debris and death.

To make everything worse, Putin had assigned the task of capturing the city to the "Wagner" mercenary group, led by their notorious commander, Yevgeny Prigozhin. The very thought sent chills down my spine. Prigozhin, a powerful oligarch, had recruited most of his fifty thousand troops directly from Russian prisons and rehabilitation camps. These men were hardened killers who took pleasure in their brutal work, unleashed with the sole purpose of spreading chaos and destruction across Ukraine. The mercenary group's actions had long been shrouded in secrecy, operating within the murky realm of plausible deniability. By design, their origins and organisational structure remained shadowy, more akin to "black ops" than a conventional fighting force.

Whilst I had already seen more than I had bargained for during my visits to Ukraine over the past two years, morgues full of burnt bodies, cruise missiles, and even trench warfare up close; the thought of coming into contact with the Wagner Group frankly terrified me. They were hardened killers, many of whom had joined up purely for the opportunity to kill without punishment. It was madness. The Russians openly distributed videos of Yevgeny Prigozhin, their mercenary leader turned rebel commander, touring prisons across Russia, offering convicted murderers, rapists, and the deranged their freedom in exchange for six months of service. Disturbingly, it became clear that he was taking on the stature of a cult leader. Many of those prisoners chose to remain even after their time was served, preferring to fight without restraint rather than return to ordinary society.

In response, Ukraine defended the city with a combination of conscripts bolstered by regular units from the National Guard and the Kalinoŭski Regiment. Leading the operation was Oleksandr Syrskyi, a charismatic and capable commander whose leadership inspired his troops as the city faced the grim reality of encirclement and defeat.

Everyone in Kyiv understood that the city had become a critical focal point in the war. Some military analysts were already calling it the bloodiest battlefield since the end of World War II, with fighting described as nothing short of "intense."

Regardless of the Russian advance, the city currently remained Ukrainian, and the people of the region were determined to defend it at all costs. Bakhmut was rapidly emerging as one of the pivotal battles of the "Russo-Ukrainian" war, carrying immense symbolic significance for both sides. From my perspective and that of those on the ground, the conflict was described as a "meat grinder" and a "vortex" devouring soldiers from both the Ukrainian and Russian military. The ferocity of the fighting and the staggering casualties drew comparisons to the Battle of Verdun in World War I and the Battle of Stalingrad in World War II.

Operational Preparation

By the time we returned to Kyiv and had rested for a couple of days, Bakhmut dominated every conversation in the hostel and the warehouse, what was happening there, and how we could possibly help. With a team already on the ground, I felt deeply invested, weighed down by guilt. How could I even think about going home without at least visiting the city and supporting the Australians? I felt trapped in an impossible situation. The only answer was clear; I needed to reach Bakhmut as quickly as possible and get them out.

Without over-thinking, I reached out to Ella and Ned, asking them to compile a list of what they needed most. I explained that we were on our way with supplies, and to evacuate them. To my amazement, after weeks of bitter complaints about the conditions, their first response was flat refusal;

"No. Why would we leave in the middle of the battle?"

We had only hours before the convoy was due to depart, leaving me little room for debate. I chose not to confront them directly. Instead, I replied that we were coming regardless and would talk it through when we arrived, but they had to be ready to move if the situation demanded it.

In the background, and behind their backs, I also spoke with Max about organising Ella and Ned's evacuation. I reached out directly to the team's local battalion commander, stressing the urgency of evacuating the Australians. It was clear that losing two aid workers, especially after they had been offered evacuation just weeks before the fall of the city, would have serious political ramifications.

Both Max and the battalion commander were impressed by the fact that we already had people on the ground and grateful for their many months of service. They immediately sprang into action, leveraging their contacts in Kyiv and beyond to secure additional equipment, vehicles, and critical medical supplies.

Through our existing network, support started to pour in from organisations like UK Aid and Ukraine Patriot, as well as major international agencies such as the Red Cross and UK-Med. Within days, we had accumulated enough supplies to fill eight or nine trucks. The only obstacle left was manpower. With such a large convoy, the biggest challenge now was finding enough drivers willing to take the risk and make the operation a reality.

It had been years since I last drove a truck, and I could barely remember how to shift gears when hauling a heavy load. However, I had been driving an ambulance in Kyiv for months, something I'd never done before. Without overthinking it, I signed up, convincing myself it would be fun and if nothing else, character-building.

Feedback from the Battalion commander in the Bakhmut Oblast was dire as most of the city was now encircled with fighting in many of the suburbs.   Understanding the local situation was essential, knowing the conditions made us understand that we were not going to be able to deliver the aid directly in to the city and would need to stop at a safe distance outside.

This news in itself made me shudder and reassess and question the sanity of arranging such a significant aid run at this point in the battle. Sat quietly one evening in the hostel I spoke to Max raising my concerns, highlighting the "intensity" of the fighting and the risk of artillery fire on the convoy.

Max was a Ukrainian born native who had been educated in the West and spoke passionately about his desire to complete the delivery, he assured me that every opportunity was being made to keep the main vehicles out of artillery range, transferring the shipment in to high speed four wheel drive vehicles that would shuttle everything into the city bringing out refugees and the injured over a period of four or days depending on conditions.

Even though I had given myself time to recover, I was reluctant to continue with the project, if I had been honest with myself, after visiting Kherson and witnessing the horrors of the front-line, I was scared. Frightened, like a small child cowering in the corner, hoping that the "monsters would go away".

Understandably, Max remained committed to the project and keen to make the most of the momentum that had built up and availability of aid. With the input of the UK Aid, Red Cross and UK Med, there was no backing out, there was one thing for sure, the way that the battle of Bakhmut was going, this was going to be one of the last aid runs and opportunities for evacuation. We needed to get most of the remaining civilians and children out.

That night, I went to bed but couldn't sleep. Restlessness clung to me, and I woke in the middle of the night drenched in a cold sweat, haunted by images of the children's burnt bodies in Bucha. I knew that the pressure of the Bakhmut mission was now pushing me over the edge, from stress to distress. The only way forward was to get on with the job before people started backing out once they realised how dangerous the mission was going to be.

Unlike Kherson, more people had began gathering at the hostel and warehouse due to the size and urgency of the conditions, ready to join the team. While they seemed friendly enough, with well-established credentials, I only knew the team leaders and the original people from the hostel, the whole situation only heightened

my unease.

Apart from Max and the original members of the team, I had come to know many of the team members by their call signs or code names, normal operating procedure in our circumstances as it provided us with a layer of protection for both our selves and family. Any intercepted communications would not reveal who we were.

CHAPTER 13

Bakhmut

All the paperwork was complete, and we were ready to go. Max and a couple of the team came to my room just after midnight. Like me, they were overwhelmed with anticipation, unable to sleep, and eager to get to the warehouse to start loading supplies.

By the time we reached the loading dock, it was still dark and bitterly cold. We needed multiple cups of coffee and sandwiches just to get ourselves moving. Despite the team's motivation, there was a quiet, tense anticipation hanging in the air.

Before setting off, we gathered the team once more to walk them through the operation. On a large notice board, we detailed the full plan, timings, the exact offloading location, and how the aid would be repacked into smaller, more agile vehicles just outside the city. These would then move swiftly in batches to designated distribution points throughout the city and to various hospitals. Each target location had been marked in colour-coded zones on a map for clarity.

The convoy had grown so large that I'd planned it like a military operation. Everyone had call signs, and we were on a strict schedule. Radios and phones were to remain off and only switched on for five minutes at the top of each hour, to reduce the risk of Russian surveillance and targeting.

As Max and I had been the main instigators of the mission— together with one of our team members, Marianna, our call signs were drawn from the national bird of Ukraine, the Nightingale (Соловей, Solovey). Max was given the honour of Nightingale One, Marianna was Nightingale Two, and I was Nightingale Three. The other designations followed the theme of birds: Owl (Сова, Sova), Hawk (Яструб, Yastrub), and Swan (Лебідь, Lebid).

When I read the briefing aloud and Max translated where necessary, I noticed both him and Marianna smiling gently as their designations were announced. They had earned it. Although I had spent months planning the mission with him and securing much of the aid from overseas, it was Max who had been the man on the ground, coordinating vehicles, recruiting the rest of the team, and managing the battalions and troops we worked alongside. What had begun as a modest effort with only a few vehicles had grown into a convoy of several, carrying hundreds of tonnes of aid and medical supplies.

As we handed out information packs to the team, I could sense their unease, especially with the change in conditions and the deterioration of the position in the city. I knew they needed reassurance, and I took the time to explain why every detail mattered.

That's when Borysko, Hawk 3, (Яструб), one of our Ukrainian volunteers, spoke up. He voiced concerns about the timing, specifically our plan to break up the convoy during the final hour before reaching the offload point. This meant increasing the spacing between vehicles to at least five minutes apart, or around two kilometres.

What worried me more than his question was the way he asked it, his voice trembled, and he looked genuinely terrified.

Max and I exchanged a glance. The team was clearly unsettled. I responded calmly, explaining that the staggered approach was designed to lower the risk of an aerial strike and to give the local military a better chance to support us as we moved through exposed countryside.

It was a good answer, reassuring, practical. But even then, I could feel doubt creeping in. Not just among the team, but within myself. We needed to get moving before more doubts surfaced.

We finished loading the lorries under the cover of darkness, before everyone else was awake trying to keep the operation as covert as possible and before long we found ourselves in convoy heading south as twilight began to show itself.

Most of the aid had been packed so tightly into the trucks that not a single extra box could be squeezed in. With a solid rapport and a shared sense of purpose, Max and I sat together in the lead lorry as we pulled out of Kyiv, both of us quietly contemplating the long road ahead.

It had been over a decade since I'd driven anything larger than a five-tonne truck, and it showed. For the first hour, I struggled to shift gears smoothly, fighting with the clutch at every stop. But determination has a way of compensating for rustiness, muscle memory kicked in, and gradually, the process became less clunky. The more I did it, the easier it became. We just needed to keep moving forward.

Our convoy now consisted of over fifteen volunteers spread across six large lorries, a smaller van, a four-wheel-drive recovery Land Rover, and our ambulance. It didn't take long for the recovery vehicle to prove its worth, especially as the terrain began to break down.

As we pushed east toward the combat zone, the roads worsened dramatically. Potholes turned into shell craters. Checkpoints multiplied, many fortified with concrete dragon's teeth and sandbags. The deeper we went, the more difficult access became. Progress slowed to a crawl as we were forced to stop repeatedly, sometimes to reroute, other times to physically clear obstacles with the recovery vehicle.

Still, the response from the local guards manning the checkpoints was largely supportive. Seeing the sheer volume of aid and the medical insignia on our vehicles, they helped us through without issue. But unlike Kherson, the local military looked visibly defeated. Tired eyes, hollow expressions, men who had seen too much and had too little left to give.

We kept pushing on, hour after hour, fuelled by coffee and momentum alone.

Around the halfway mark, a new vehicle joined the convoy, clearly marked with military insignia. A liaison officer from a battalion stationed near Bakhmut. While I knew that a liaison officer would be accompanying us at some stage, no one had mentioned an escort vehicle. I turned to Max, unsettled, asking him, "Why did we need a military escort vehicle when our vehicles were clearly marked as medical, this only made us a legitimate target."

I was certain he'd known all along what I was going to ask and had hidden the truth from me deliberately because of what it implied.

Yes, I was angry with him, and he knew it. Yet it had always been difficult to stay angry at Max. As he tried to reassure me, the confidence I usually saw in him, that easy bravado, was gone. His silence spoke louder than words ever could. The longer I thought about it, the clearer it became: the Russians didn't care. Whether it was a convoy of civilians, ambulances, or aid trucks, the presence or absence of a military escort meant nothing. They would target anyone, regardless.

As the day progressed, the skies cleared towards the east as the sun moved overhead and the distant outline of the city came into view. Clouds of smoke appeared, drifting slowly across the horizon. Max and I instinctively looked at each other.

Bakhmut was on fire from one side to the other, quietly smouldering in the distance. The closer we got, the more smoke and pyrocumulus clouds appeared from the constant artillery fire.

From where we were, the city of Bakhmut burned like a scar on the horizon. Even at such a distance, the glow of fire dominated the sky, an immense dome of shifting smoke, orange and red, flickering like the very edge of hell. Black columns of smoke rose higher than clouds, spreading outward until the entire skyline was consumed in a choking haze. The ground beneath us seemed to tremble faintly with the echo of distant detonations, a constant reminder that this was not a natural fire but the death of a city torn apart by war.

Bakhmut, city on fire

The scale of destruction was beyond anything my mind could process. I could see no defined outlines of buildings, no familiar signs of life, only shifting shadows swallowed by flame. The thought that people once lived there, that streets once bustled with voices and laughter, was unbearable against the infernal reality before me. It was not a city anymore but a furnace, a place erased by violence.

I felt frozen in time, unable to process the image, unable even to summon the words that might describe what I was witnessing. Any attempt felt hollow, inadequate against the enormity of the sight. My eyes burned, not just from the acrid smoke carried on the wind, but from the shock of realising that I was looking at the complete destruction of an entire community. Bakhmut was no longer a place on the map, it was a warning, an image of absolute ruin carved into the horizon.

An hour away on the outskirts of the city, we were directed off the main road and onto a short gravel track leading to the edge of a forest.

In anticipation of our arrival, we were directed in to prepared parking locations, and the convoy scattered into cover under the trees and evening mist. Within minutes, people began to appear from camouflaged huts to cover the lorries with nets. As quickly as we could get out of the vehicles, they had been hidden out of sight.

The stark change in atmosphere had an immediate impact on us. We could instantly see the peoples need for aid and gratitude. Bakhmut had been under siege for months and was on the edge of being overwhelmed.

As quickly as the people had appeared, other smaller cars and vans had moved into place ready to begin the transfer of equipment. We were keen to ensure the smooth delivery of aid and attempted to take control of the distribution, calmly refusing to open the back of the lorries until we had connected with the company commander as planned.

Full of anticipation, the locals remained patient as we arranged the distribution of supplies and directed the army to open the first two vehicles, disbursing their contents into batches as the evening progressed.

For the time being, we left the main convoy behind and transferred the aid into smaller, more agile vehicles. Following the carefully devised plan, we prioritised medical supplies, unpacking each vehicle one by one and setting aside most of the food for the following day. The only vehicle from the original convoy to continue in was our ambulance, driven by Max and me.

Once the first lorries had been unloaded, we stationed a guard by the remaining aid and headed toward the city with a "heightened" sense of anticipation. Driving at night was challenging at the best of times and nearly impossible without headlights. However, the local "fixers" possessed an instinctive knowledge of the ground and route into the city. A light dusting of ash and concrete dust reflected the moon light off the road helping us to navigate every twist and turn.

Moving under the cover of darkness made sense. The smaller vehicles were not only more manoeuvrable but also made difficult targets, far less likely to attract attention. This allowed us to advance with purpose. Yet, the closer we got, the unmistakable sound of gunfire and "explosions" grew louder, the battle's intensity was unmistakable.

By now, our team had been awake for over thirty-six hours, the exhaustion etched deep into our faces. Even Max had begun voicing his concerns, his frustration justified as he stumbled over words and our vision began to blur. We were teetering on the edge, our bodies threatening to shut down or collapse from sheer fatigue.

Seeing our distress, the new liaison officer insisted that we stopped to rest. Whilst his insistence was both practical and compassionate, the first batch of aid needed to reach the central hospital under supervision before dawn. After some discussion, we reluctantly agreed to rest once we reached the hospital and continued to drive onward.

Shortly after, we pulled up to what remained of the hospital. Most of the upper floors had been "obliterated", leaving only the shattered remains of a building above. It was almost impossible to comprehend how this place had ever functioned as a medical facility.

Knowing we were on our way, parking bays had been cleared and rubble pushed aside in anticipation. As our small convoy started to arrive rolling to a stop, the silence of the night broke, the first figures appeared, medical staff and porters emerging cautiously from the basement, blinking against the cold air. Within moments, they had arranged themselves with quiet efficiency, forming a human chain without a word. Relief parcels passed swiftly from hand to another, moving steadily into the safety of the underground shelter.

Time itself seemed compressed; every second outside carried its own risk. The urgency was immediate, no one lingered, no one hesitated. Out here, exposed in the open, danger pressed in on all sides, and the unspoken understanding bound us all: the faster we moved, the more lives might be saved.

Once unpacked, the staff could see us struggling to remain upright, with our limbs trembling from sheer exhaustion. Still, their gratitude cut through the haze of fatigue as they thanked us for our courage and humanity. In just over twenty minutes, all the packages had been unloaded and the people dispersed as we followed the others inside, desperate to find a quiet place to sleep.

Amid the darkened hallways and rubble-strewn corridors, we found enough space and quiet corner in an abandoned storeroom to make camp. The floor was cold and unyielding, the air thick with dust and silence. The walls, scarred by time and violence, whispered of emptiness and death. It was far from a place of rest, but drained by exhaustion, we surrendered to sleep almost instantly.

The team had arrived and we had survived the first objective.

Under Fire

By morning, the hospital staff had anticipated our disorientation and kindly prepared a hot pot of coffee. We gratefully devoured it alongside the protein bars we had brought with us.

In the distance, the sound of artillery fire echoed, a grim reminder of the war unfolding around us. As we sat quietly, eating breakfast, we tried to come to terms with our new reality. There was little time to linger. Max and the others were already packing up, signalling the start of another day.

The team now split into groups of two or three, depending on their assignments. The golden rule was clear: no one was ever to be left alone or without communications. The first two teams remained at the hospital, shuttling refugees and wounded back to the evacuation point where the larger vans and lorries waited, before loading additional supplies for the return journey into the city. The third team, made up of a trained medic and a fixer, was tasked with visiting another medical site. Meanwhile, Max and I had an appointment later in the day with the battalion commander to discuss the unit's needs for the coming weeks after our first run out of the city. Having learnt from my mistake in Kherson, Max had promised not to raise my interest in drones; I had no wish to be dragged into another trench-line demonstration of troops hurling grenades at the enemy.

Within twenty minutes, we were taken to a recovery station where the walking wounded and selected refugees had gathered for evacuation. They were a weary, desperate group, weak, filthy, and understandably irritable. Most carried only a small bag or a few plastic shopping bags filled with the bare essentials. Many were injured and needed to be transported to a more advanced medical facility in Kyiv.

Most of the city lay abandoned, its civilians already forced out. Those who had chosen to remain were beginning to grasp the hopelessness of that decision and had packed their belongings, ready at last to leave.

The vehicles from the previous day had been hidden away under a light dusting of ash and rubble, kept out of sight. Time was against us. We needed to move as many of them as possible before dawn. The cold dusty conditions had taken their toll on the engines, making them stubborn to start. At least two vehicles required a jump-start, an expected struggle due to the conditions.

Moving the refugees before daylight was crucial. Darkness offered some protection, making the journey safer. As each vehicle warmed up, it was positioned for departure. Groups of wounded were loaded and sent out of the city in carefully timed intervals, one car at a time, with at least ten minutes between each departure to avoid drawing unwanted attention.

We were operating on a prearranged schedule, one that had been meticulously planned in advance. That structure, however fragile, gave us a small but invaluable sense of purpose and reassurance.

All the guards on the city's outskirts recognised us and waved us through with the designated signal. Thanks to our well-planned agreement, reaching the camp in the amber zone took only twenty to thirty minutes.

Arriving at our destination felt more familiar than I had expected. The atmosphere in the camp had improved, and those awaiting us were ready with hot coffee and a warm space for the refugees to rest before their journey continued. While we had been travelling, the remaining vehicles had already been unpacked in preparation for our arrival.

Everything moved like clockwork. The refugees were unloaded as efficiently as the new aid packages were packed back into the ambulance. Refreshed and fuelled by coffee, we set off once again, heading back towards the city centre forward operating base to meet the commander.

Each mile back toward the city brought us one step closer to completing our mission, filling me with a cautious hope that we would reach our final goal. But as we neared the next checkpoint, peering through the morning mist, the ground suddenly erupted.

Artillery shells crashed down just meters from our path. I had momentarily forgotten, we were now deep in the red zone, well within artillery and drone range.

The force of the explosions shook the vehicle. A sharp crack spread across the windshield as I froze, my breath caught in my throat, my hands locked around the steering wheel. Max's eyes fixed onto the horizon. The unmistakable shadow of a drone loomed in the distance. We both knew instantly, we were in deep trouble.

My first instinct was to slam on the brakes, nearly sending the vehicle skidding off the road. But the realisation hit me almost immediately, we couldn't stop. We had to move faster, stagger our movements, make ourselves a harder target to track. Smoke and debris filled the air, obscuring my vision as fear surged through me, tightening my chest and stiffening my limbs.

Max understood exactly what was happening. His voice cut through the chaos as he began counting aloud, anticipating the next artillery strike. If only one position was firing at us, we had fifty seconds before the next salvo.

I spontaneously began to weave down the road changing speeds as my training kicked in. Seconds seemed to become minutes as time froze driving towards the next checkpoint just visible in the distance. Seeing that we were in distress some of the guards that had not taken cover began to fire at the drone above them hoping to distract the incoming artillery.

Chaos surrounded us. I suddenly became aware that I had been holding my breath and gasped for air just as another salvo slammed into the ground behind us. The explosion shattered a nearby window, sending shrapnel slicing through the aid boxes stacked behind us. Instinctively, Max and I ducked, our eyes locking as we checked each other for injuries. Wide-eyed and trembling, we recognised our luck; we had been spared. Without hesitation, we surged forward, pushing faster toward cover.

By the time we reached the checkpoint, adrenaline and fear had taken hold, leaving us shaking as we leapt from the vehicle and dashed into the nearest bunkers. Soldiers ushered us inside, and we collapsed behind reinforced barricades and trench walls, staring at each other in stunned silence. The Ukrainian voices around us blurred into an indecipherable hum. I turned to Max for guidance, his eyes tracking the guards' conversation. Then, looking directly at me, speaking first trying to find out whether I was hurt. For a second,

I wasn't sure. My hands skimmed over my arms and legs, searching for any sign of blood.

Barely able to speak, I stammered, "No, I'm good. You?"

Max exhaled sharply, then cracked a shaky grin. "No, I'm good too mate, but I think I just shit myself."

Laughter bubbled up between us, a brief release from the terror still gripping our bodies as we hugged each other in gratitude. Another artillery barrage thundered outside, but we were safe now, concealed within the tree-line and the bunker complex on the outskirts of the city. Only then did we take stock of what had happened.

The ambulance had taken the brunt of the blast, its armoured side panels scarred by shrapnel. The reinforced plating had done its job, absorbing most of the impact, but the rear tyre hadn't been as lucky, it had been shredded, now lying flat against the dirt. I touched my forehead and winced, I had scraped it while scrambling out of the car. Just a scratch. Nothing serious. More than anything, I was just relieved that both of us were alive.

The damaged ambulance

Still shaking, I decided to stay put for a moment, letting the fear subside until I could breath again. I rummaged through the ambulance, pulling out a pack of biscuits and a tin of tea. If there was ever a time for a strong brew, it was now. The soldiers around us seemed unfazed, treating the shelling as just another part of their day. Sitting with them, sharing tea and broken conversations, the situation slowly began to feel more manageable.

Our team radioed ahead to warn the next vehicle about the incoming fire, giving them the chance to delay their approach. But they made the call to press on. From our cover, we watched them weave their way down the road, dodging potholes and debris, their determination unwavering.

As they rolled up beside us, we stepped out to change the tyre. Despite the chaos, there was an undeniable sense of triumph, we had made it through unscathed. The approaching team slowed, rolling their windows down to exchange grins, thumbs-ups, and a few well-earned laughs before pushing on. We waved them through, knowing we'd meet again soon to unload supplies. Another story added to the growing collection.

And as surreal as it all felt, one truth settled in: we weren't just passing through danger anymore. We were targets.

Arriving in the relative safety of the city centre and hospital district should have been reassuring. Marked as a hospital, there was hope it would be spared from Russian attacks. But deep down, we all knew that was far from reality. With much of the city's aid distribution centred around the hospital and its surroundings, it had become a prime target. There was no sense of humanity in the situation, everything we had brought to help civilians and the wounded had instead become a legitimate target. The Russians showed no mercy, no compassion for the young, the families, the sick, or the injured.

The Australian Team

Meeting Ned and Ella was a relief. Though I had spoken to them for months and had studied their personnel files in detail, nothing could replace meeting them face-to-face. Both were young, with limited life experience, yet they had pushed beyond that by dedicating themselves to service, working with the defence force and in high-risk locations like Ukraine. Their stories from the eastern front were staggering, and I was quickly overwhelmed by the sheer number of casualties.

Ella described going out on patrol on multiple accessions to forward aid stations, only to find that just three or four men had returned from a unit of thirty. Ned confirmed that they had been seeing such devastating losses, with grievously wounded soldiers, on a regular basis for the past six weeks. The battle for Bakhmut was beyond comprehension, sickening, inhumane. At its worst, the Russians refused to take prisoners, executing anyone who tried to surrender when they became isolated or overrun.

The battle had now moved from the salt mines of Soledar, a 'network of underground tunnels', prised by Russian mercenaries to the central industrial area of Bakhmut, AZOM. If the Russians could completely overrun the mines, they would gain hundreds of kilometres of tunnels, potentially accommodating thousands of

troops and vehicles at depth under-cover, providing a strategic advantage for movement and defence. More importantly, the area secured the northeastern flank towards Bakhmut and could disrupt supply lines to the city, which was now already under heavy assault.

Max and I sat for hours, listening to their stories compiling a list of their most urgently needed supplies and medical aid. It didn't take long for us to understand they had no intention of leaving, despite our plan to evacuate them in the next couple of days. Ella had taken shrapnel to the leg, while the wound wasn't severe, it had been troubling her for weeks. There was a strong suspicion that fragments remained embedded, causing persistent swelling and a noticeable limp. To make matters worse, Ned refused to leave her side. They both understood that, eventually, they would have to go, but not now. No matter how traumatic or dangerous the situation, they simply wouldn't walk away.

The more they spoke about their connection to the local people, the more we understood their determination to see this through. It was pointless to try and convince them otherwise.

In return, I was overwhelmed with guilt, and before long I felt the same as Max. We had been talked into accompanying them to the front lines, where we would meet another battalion and a front-line medical team. Later, I admitted to Max that I felt like an idiot, cornered into an even worse situation, facing greater danger than before, this time up close and personal with the Wagner Group.

Both Ned and Ella spoke plainly: we had to visit the troops, moving from the aid station toward the front-line trenches. The battalion commander needed us there in person, to help bolster morale by spreading the news that a major aid delivery had arrived and was on its way. For the men, it meant they would eat well for the next couple of weeks and know their wounded comrades would be evacuated. To them, it was essential.

For us, it meant risking our lives for a few days, perhaps a week. For those fighting in the city, it meant staying behind, fighting on until they were forced to die there.

Before we could come up with a way out, the remaining aid supplies were being unloaded, and the decision had already been made. Before long, we were on our way to the front-lines and trenches again and the forward operating base of the Kastuś Kalinoŭski Battalion, part of the International Legion.

It was impossible to get close to the location safely in one go by car, so we began to walk through a maze of trenches tunnels and interconnecting buildings towards the front as the sounds of war grew "louder" and "louder" around us to the point where it became difficult to hear anything else.

It was at this point that I found myself in a haze of darkness and disbelief. We had just begun to move through a vast industrial building, shattered from one end to the other. Glancing upward through gaps in the ruined roof for the threat of drones, we dashed from one heap of rubble to another, crouching for cover beneath fallen debris and massive industrial girders. Then it struck me: we were walking through a scene straight out of the film "Stalingrad".

The moment was surreal. I had watched the film many times, always struck by the way it showed the Russians as the defenders, the good guys, holding out against the German invaders. It left me with a profound sense of both loss and admiration for those who had resisted Hitler, the fascist dictator of Nazi Germany. Now, eighty years later, I found myself in a disturbingly similar situation, trudging through the ruins of a colossal industrial area in the heart of Bakhmut, standing on the side of democracy against Vladimir Putin and his authoritarian regime.

The irony was not lost on me. When we finally paused for a short rest between buildings, I smiled at Max and told him what had gone through my mind. We laughed together, quietly, at the absurd symmetry of history. Surprisingly our goal was closer than we had expected and only took us three or four hours to walk before descending into another trench entrance.

Understanding that we would need to stay overnight or for a couple of days, we had taken our armoured vests, survival gear and packs. This quickly became essential, as there was limited food and water available.

The appearance of aid workers in the trenches and aid stations around the trenches took many by surprise, the local soldiers seemed to take a double look at us as we walked past and spoke to them shaking their hands whenever possible saying, "Slava Ukraine", handing out small packets of biscuits and sugary sweets.

They were very pleased to see us and grateful that we were there, anyone who could speak English quickly asked us what we had brought for them and we quickly became acutely aware of the reasons why we had been asked to visit the trenches. We were there to improve morale, and it was beginning to work.

The more information we divulged, the happier the people around us became realising that they had been brought fresh food, water and warmer gear. The world had not forgotten them.

The Salt Mines of Soledar

At the point we reached Bakhmut, the salt mines of Soledar had already fallen, and the remnants of fighting men lay scattered between the towns like broken dolls. The city was a wasteland of ruin, where the air itself seemed thick with death. Smoke and dust clung to every surface, fires burning in the hollowed remains of buildings painted the streets in a sickly orange haze. Artillery pounded without pause, the relentless heartbeat of destruction shaking the earth and rattling my teeth. Machine guns stuttered from shattered windows, rifles cracked sporadically, and each sound carried the finality of another life taken.

Bodies lay where they had fallen, soldiers and civilians alike, frozen in grotesque poses of agony, some half-covered with tarps or coats, others exposed to the elements. Limbs jutted from rubble, hands curled as if in prayer. The stench was suffocating: rotting flesh, smoke, and cordite mixing into a poisonous fog that settled on the tongue and lungs. Every breath tasted of death.

The defenders moved through the ruins like spectres, faces smeared with soot, uniforms torn and bloodied, eyes hollow but burning with resolve. Even the wounded clung to their posts. They fought not only for Bakhmut but for the idea of resistance itself. To lose this city was to surrender hope.

An aid station had been carved into the basement of a half-collapsed building. The air inside was thick with blood and sweat, layered with the metallic tang of antiseptic and decay. Men groaned in every corner, some missing limbs, others clutching shrapnel wounds. Medics, gaunt and sleepless, worked with trembling hands, tearing shirts into bandages, rationing morphine drop by drop. When Max and I arrived, their faces turned toward us, desperate, asking if we'd brought help.

For once, we could say yes. We unloaded two heavy kit bags, pressure bandages, antibiotics, IV fluids, morphine, and their reaction was electric. They tore open the boxes like children on Christmas morning, relief and disbelief flashing across exhausted faces. Even the wounded smiled, some crying openly when they realised they might finally be going home. The bunker transformed in moments, despair shifting into gratitude, fear into fleeting joy. For the first time in days, I felt we had made a difference.

Outside, the city screamed. Shells ripped through buildings, walls collapsed into plumes of dust, and the ground trembled with every impact. Night brought no peace, only shadows cast by fires, rats skittering through the trenches, and the cries of the trapped echoing faintly through broken masonry. Drones buzzed overhead, their distant whine a constant reminder that nowhere was safe. Still, the defenders held. They reinforced what little shelter remained, bracing themselves against the next barrage.

After hours moving through the trenches, Max and I returned to the aid station carrying two newly injured men. Once they had been treated, exhaustion hit us hard. We found a quiet corner of the makeshift bunker, brewed a warm drink from what little we had, ate a few scraps, and tried to rest, sleeping if possible, or simply lying still among the rubble, the stench, the ceaseless noise of war, and the groans of the wounded. We spent the night huddled there, too drained to speak, while explosions rolled like distant thunder and the ground shivered with every detonation. By dawn, we planned to move west toward the main hospital, but renewed shelling had pinned us in place.

CHAPTER 14

The Trenches

The next day, we planned to move back toward the main hospital and evacuation point to rejoin the others, but incoming fire had continued to trap us in place, making it impossible to move the wounded. Progress was agonisingly slow; any attempt to move more then one at a time would have created a larger, more visible target, an invitation for the enemy to strike.

Many of the soldiers we met had escaped Soledar. The memory of that slaughter clung to them, a haunted knowledge that survival was only temporary. They fought on because to stop was to die, the Wagner Group were not taking many captives, preferring to kill anyone on sight, earning the reputation as, "Putin's ghost army".

The city's smell now gnawed at the back of my throat, overwhelming my senses, cordite, smoke, and the iron sting of blood layered with the sickly sweetness of decay. It clung to my clothes and hair, a punishment with every breath. Fires smouldered in shattered apartments, their smoke drifting across the skyline in lazy, suffocating sheets. The sounds were unending: artillery thundered in rhythm, mortars whined overhead, machine guns rattled from unseen corners. Even silence had been destroyed.

Among the ruins moved a few civilians, nurses, volunteers, the last to flee, faces pale and hollow, carrying what little they could salvage. Their eyes told the same story: fear, loss, determination. Each one we met, we urged westward toward safety.

We pushed deeper into the industrial zone. The AZOM complex loomed like a steel citadel, factories and silos blackened and rusted, walls scorched by phosphorus and fire. Defenders had turned every corridor into a fortress of sandbags, crates, and overturned vehicles. The city was a vast, bleeding chessboard, every square fought for in blood.

Max and I struggled to understand how we had ended up there. We had been following the guidance of our local fixer, Mykhailo, who had been assigned to us by the battalion commander with strict orders to keep us safe. Instead, we now found ourselves deeper inside the combat zone when we should have been moving the other way.

I could see Max's frustration rising as he spoke with Mykhailo, and you didn't need to understand Ukrainian to know what was being said. It was rare to see Max lose his composure, even under fire, but this time he was close to snapping.

Our local fixer, Mykhailo, insisted that he had been told to take us to see AZOM. He was following orders from his commander, who wanted us to witness what holding the line truly meant. Max argued, his patience frayed, but there was no turning back now as we needed to find hard cover. Inside. The noise was deafening. Wounded men cried out from every corner, medics worked knee-deep in blood, and explosions shook dust from the ceiling like rain. And still, they fought. Every face we passed bore the same defiance.

We now found ourselves at the mercy of Mykhailo, agreeing to move through the complex for a brief tour, handing out what little we had brought, more boiled sweets, snacks, and medical supplies. The soldiers were visibly stunned to see us there, two international aid workers with medic patches on our arms, sharing the same trenches under fire. Yet their surprise quickly turned to encouragement; our presence seemed to remind them that someone, somewhere, still cared. Many asked again what we had brought, seeking more reassurance.

Minutes blurred into hours as we darted between trenches, searching for cover from the constant bombardment. By early afternoon, drained and shaken, we insisted on returning to the relative safety of the main aid station. The simple act of moving through that environment was soul-destroying. Every shattered wall, every broken body, etched itself into memory. I saw the same strain in Max's eyes, we were both at our limit.

Heading west from the AZOM complex took far longer than expected. Each time we thought it safe to move, another burst of fire or an artillery round would land close, pinning us in place. When the shelling paused, we ran, weaving between positions and exchanging brief nods with the soldiers we had just met. Their faces brightened when they recognised us again, and soon they began pressing small scraps of paper into our hands, notes for their families, written in pencil or bloodstained ink. Most messages were simple: "Please let them know I am still alive."

All we could do was take the notes and promise to deliver them when we returned to Kyiv. Each farewell ended the same way, with a quiet "Slava Ukraini" and the shared hope that their nightmare would soon end.

Unlike Kherson, Bakhmut was overwhelmed with a sense of defeat. Most of the soldiers and civilians trying to escape the carnage already knew the city was lost, much of it had been encircled. Yet amid the despair, there were still those who remained. A dedicated contingent of troops had chosen to stay, and even in the face of near-impossible odds, their resolve never wavered. Looking at the casualty figures, it was almost incomprehensible. How could the majority of the troops sent out be wiped out, eradicated, only to be followed by another group, and then another? The scale of loss was bizarre, unfathomable, yet these men and women stayed because this was their home. The majority of Bakhmut's people had wanted to fight, to defend their city, their land, their freedom. Their dedication to Ukraine was staggering. No one had to tell them to hold their ground; they simply would not leave.

By the time it was dark, it was too late to move any further, we had been overcome by fatigue and just needed to rest. That night, Mykhailo sat with us beneath a fractured beam. He spoke softly of God and the reverence owed to the fallen. Then, with quiet gravity, he asked if I would speak words over the dead the next day. The request struck deep. I nodded, unable to speak through the weight in my chest. Then he showed us a video on his phone, the Russians dumping their own dead into a mass grave with a digger. Bodies poured from a truck like refuse. Within seconds, bile rose in my throat. I turned away wrenching unable to vomit. Those men had fought and died, and their bodies were discarded like waste. The horror of it branded itself into my mind.

Sleep was impossible. The bombardment continued, each blast shaking the walls, each scream tearing through what little composure remained. I lay awake for another night, haunted by the thought of how long a man could endure such a place before losing his humanity. Bakhmut burned everything away, fear, compassion, even grief, until only determination remained.

AZOM, the Last Stand

The sun began to rise. Max and I had found enough cover to heat a cup of coffee and share a ration pack. There was no dignity in the conditions we found ourselves in. We chose safety over comfort, digging a small hole with a trowel to relieve ourselves before using a disinfectant wipe to clean up. Even so, it was almost impossible to remove the ground-in dirt and concrete dust from our hair. It would be another week before we could properly wash, but the memories of what we had seen would never leave us, nor wash off.

Bakhmut was already a city stripped of its humanity. The roads were a gauntlet of destruction: charred vehicles burned in shallow craters, their twisted frames jutting from the earth like the bones of giants. The asphalt had been torn apart, pocked with shell holes and lined with shattered concrete. It was becoming clear to many of us that AZOM would be the last stand for the soldiers we had just met, men who, in all likelihood, would never be seen again.

By now, Max and I were seriously concerned. We had been out of touch with our team for over thirty-six hours, forced on a detour to witness some of the worst fighting Ukraine could endure. Still, we understood why the battalion commander had wanted us to see the remains of the city and the conditions in AZOM. It made clear what they needed to defend the district and maintain their foothold. With that in mind, we sat quietly with Mykhailo, discussing our next steps, making notes for the commander, and trying to anticipate what questions he might ask when we returned.

In the distance, between us and the aid station, smoke rolled in lazy curtains, curling above shattered rooftops and gutted apartment blocks. The sun, if it had ever shone bright here, was now veiled by haze, casting a dull, sickly light over the streets. Fires burned in random places: overturned cars, collapsed buildings, ruptured gas lines. The acrid tang of burning fuel and plastic mixed with the stench of metal, forming a chemical chorus that assaulted the senses with every step.

The scale of destruction was beyond comprehension. Each step forward felt like a negotiation with death itself. I tried desperately to look away from one scene, a rat tearing at the corpse of a young man, but the image seared itself into my mind. My stomach lurched; acid burned my throat before erupting in a fountain of vomit, tears streaming down my face.

Max's reaction wasn't far behind. He turned, caught sight of the same horror, and instinctively pulled me under his arm. He urged me forward, guiding me toward cover as we stumbled on with our escort. Shame hit me almost as hard as the nausea. I muttered an apology in the few Ukrainian words I knew: "Вибачте." Max gave a faint nod of understanding. Moments later, in a fragile moment of compassion, we embraced.

Our young escort, Mykhailo, tried to reassure us, admitting that he too had been overwhelmed, that he'd vomited up more food than he could ever keep down. Trying to lighten the mood, he pointed to the body and said we shouldn't worry, it had been a Russian, not a Ukrainian, otherwise it would have been recovered sooner. "They deserve to be left for the rats," he added matter-of-factly. It was a brutal sentiment, but his way of easing the tension. Stunned by the comment, Max and I looked at each other before breaking into uncontrollable laughter, helpless, cathartic, absurd.

It was at that moment I knew I was broken. I needed to leave before I lost what remained of my own humanity and composure. Max and I had seen the worst the war could offer. Now we understood why the battalion commander had wanted us to witness AZOM and the trenches for ourselves. There was no doubt, we had seen the depths of inhumanity Russia had unleashed upon Ukraine and its people.

Mykhailo sensed the change in us. We insisted we had seen enough and that it was time to go back.

We moved with purpose westward, heading toward the primary aid station where Mykhailo had been instructed to escort us. As we passed through the ruins, we urged the wounded to come with us, slipping through shattered streets, clutching their equipment or dragging whatever meagre possessions they could carry. Their eyes were hollow, faces pale and smeared with soot. Some sheltered in basements, others huddled inside half-collapsed buildings.

At every opportunity, we told them to keep moving toward the evacuation point we had set up in the west of the city. Each step was deliberate, chosen to avoid the streets where artillery and gunfire were heaviest. Every face told the same story: fear, loss, and a grim determination to survive.

"Move. Keep moving west, toward safety, toward evacuation."

There was no rest, no pause, no mercy. Every moment brought new casualties, new deaths, new acts of desperation. The longer I stayed in Bakhmut, the more I felt myself slipping into a timeless blur, an unimaginable abyss.

Max and I had resolved to stay only as long as necessary to deliver aid. Our mission was clear: improve morale, supply the wounded, and help evacuate as many as possible before the city was overrun. The thought of leaving was compelling, but with so many looking to us for guidance, we steeled ourselves. Together, we would endure, supporting each other, holding the line, and facing the chaos head-on

Time to Leave

After hours of moving through the ruins of the city, we finally reached the primary aid station a few kilometres back from the front line. We were accompanied by Sergeant Mykhailo and a number of wounded soldiers, many so badly injured they could barely walk, slowing our progress significantly. Though Max and I were exhausted, we were determined to return to the hospital and main evacuation point, knowing the rest of our team would be worried. We had now been out of contact for more than forty eight hours.

I wanted to leave immediately, but the battalion commander had been waiting for us, intent on debriefing before we moved on. After a short exchange with Mykhailo, Max and I noticed the shift in his posture: he understood what we had seen and would carry the report forward. Bakhmut needed more, more support, ammunition, men, supplies, yet everyone already knew the truth: the city was lost.

For five minutes, Max, Mykhailo, and the commander spoke quietly, occasionally drawing me into the conversation. The tone was heavy, sombre, marked by shared knowledge that required no explanation. They were deeply grateful for our visit and the scale of the aid delivery, urging us to pass on their message to others. To me, it felt strange, everyone already knew about the situation in the city. That had been the very reason for our mission and the urgency of the aid in the first place.

Now our priority was clear: evacuate as many of the wounded as possible, along with the few refugees still remaining, moving them beyond artillery range to safety. The medics offered us a place to sleep another night, but we were desperate to reconnect with our team. We thanked everyone, shaking hands, forcing a grin as we said "Slava Ukraini," promising that the rest of the wounded would be moved to the hospital before we finally departed.

The past three days had been a baptism of fire, something I would never fully understand, cope with, or convey to anyone outside of it. The only people who could truly understand were those still fighting in the city, and our teammates who had endured it alongside us.

Even though it was already late in the day, we left. Our next destination: the shattered remains of the hospital, still thirty minute away on foot.

CHAPTER 15

Loss

Walking back toward the hospital was a relief so sharp it almost hurt. Just knowing we would not have to return to the front-line, where every step was a struggle for life, every moment an indignity endured, brought a wave of comfort over us. We darted between ruined buildings, across rubble, from one patch of cover to another, carried by the fragile hope that we were finally safe.

The further we moved from the carnage, the quieter it became. The thunder of shelling and the crack of gunfire began to fade into the distance, and only then did I realise how deafening, how overwhelming it had all been. My body ached from the constant tension. The only sound that still clawed at our nerves was the whining buzz of drones overhead, flickering from one side of the street to the other. Soldiers we passed waved yellow and blue ribbons in reassurance, but no gesture could quiet the gut-wrenching dread that came with that sound.

By the time we staggered to the hospital, caked in mud, concrete dust, and sweat, dusk had already fallen. Rounding the corner, we gave the checkpoint guards a tired nod, sharing that faint, knowing smile of men who had survived. For the first time in days, I saw Max smile, just a flicker of joy at being alive. Relief washed over us like a tide.

It didn't last.

Apart from Max and a few of the original team, I knew most of the others only by their call signs. It was safer that way, for us, for our families. No names in intercepted communications, only birds and shadows.

With immense relief we broke into a run for the last thirty metres, grinning as we searched for familiar faces among the crowd of civilians and soldiers milling around the hospital doors. But as we reached the entrance, everything stopped.

Max asked the guard a question in Ukrainian, stumbling over his words. The man's reply was sharp, and Max crumpled to his knees as if struck.

My first instinct was to grab him, certain he had been hurt. But when I pulled him up, I saw tears in his eyes. His voice shook as he forced out the words.

Borysko, "Hawk Three", was dead.

The word hit me like a shell. "Dead". After everything we had endured, everything we had survived together. Borysko. How?

We sat together on a broken wall outside the hospital sobbing as Max, explained. Borysko had been caught in the open during an artillery strike, cut down in almost the same place we had been ambushed in the ambulance.

Darkness settled, and the stone beneath us became even colder. The weight of it pressed into our bones as the truth sank in. One of us was gone. One of our team was dead. We had lost Borysko.

Moving on

Minutes seemed to blur into hours as we sat on the wall, trying to process the news, dreading what we would say to the rest of the team. We were already overdue, out of contact for days, and now faced with the loss of one of our own.

Max sat in silence, hollowed by shock. This was the first death he had faced at the front. He had already buried members of his family during the war, but this was different, this was a brother in arms. His deepest fear was how to explain it to Borysko's family, and how to hold himself together in front of the others.

For me, instinct took over. Years of army training had drilled into me how to face reality, suppress grief, and stay locked on the mission. And the mission now was clear: hold the team together and evacuate nearly a hundred refugees and wounded. In the moment, I knew that discipline would carry me through, that I could push aside the weight of loss and keep moving. But I also knew the cost, that the raw humanity I had buried would return later, punishing me in the darker hours with nightmares

Ironically, one of my first friends in Ukraine, Andrew from the Irish Medical Mission, had once given me a recovery pack with two body bags tucked discreetly among our gear. I had prayed I'd never need them. Now, grimly, we did. With two of our team still inside the hospital, I laid out the plan to Max. We would take Borysko's body back to Kyiv for his family. Perhaps he had already been buried, but we needed to know. Max's face told me everything, revulsion at the thought of moving him, but also the hard acceptance that it was the right thing to do. He had to come home with us.

In my world, you never left someone behind. Even if he or she should have been laid to rest in Bakhmut among the other fallen souls, we would carry them home.

We wiped our faces, forcing ourselves into composure, and went inside to find Aleksy and David. The moment they saw us, I recognised the grief in their eyes. Before they could speak, we embraced them. We already knew. Then, steadying myself, I asked where Borysko lay. Their voices came together, subdued and strained. He was still out by the checkpoint, exactly where he had fallen. They had waited for us with the ambulance before burial.

Aleksy added they had boarded up the shattered window as best they could and that our ambulance was still drive-able parked outside under cover. Both of the young men were crushed by the loss, talking to us with their heads down, but also relieved to see us alive. That alone seemed to steady them. With that in mind, we sent an encoded message to the others, letting them know we were safe and on our way back.

Exhausted, desperate for even a moment of stillness, we shared a quiet cup of tea together. Then we climbed into the battered ambulance, said our goodbyes to the hospital staff, and turned toward the checkpoint. It was safer to move after dark, though each of us was bone-weary, running on empty. But there was no choice. We had to leave, get to the wounded and the refugees past the checkpoint, and rejoin the others waiting at the forest camp, ready to begin the long journey back to Kyiv.

Borysko

It was late in the evening by the time we reached the checkpoint. Driving the ambulance at night had been harder than I expected. With so little moonlight on the road, it was nearly impossible to go faster than twenty kilometres an hour without risking a crash that could disable or destroy the vehicle we so desperately needed intact.

The closer we came, the heavier the anxiety pressed on me. The thought of seeing Borysko's body churned my stomach until I felt faint. I recognised the symptoms I, had been here before. We were all teetering on the edge of shock. I forced myself to steady my breathing, to focus on the job at hand. We had to keep moving. If we allowed ourselves to dwell on the loss, grief would overwhelm us.

Pulling up to the makeshift hut and concrete bunker, the guards recognised us from a distance. We gave the correct torch signal and challenge before approaching slowly with our liaison officer.

Emma and Lucus, the last of the Hawk team, stood in the shadows. Emma sobbed uncontrollably as Lucus tried to hold her together. There was little to say. We gave them both a long, wordless embrace, reassuring them only that Borysko would come home with us, back to his family.

Like Max earlier, they recoiled at the idea of moving him. When I asked Emma where he was, she nodded toward a distant bundle, blankets and plastic, with his legs just visible in the dark.

Lucus quickly explained what had happened. Borysko had stepped out of their vehicle, just behind the bunker, when an artillery salvo struck. Shrapnel had torn through his torso in seconds. The more he spoke, the more his voice broke, until I put my arm around his shoulder to steady him. "There was nothing anyone could have done," I told him. We had done everything possible to avoid casualties, but the Russians had already been pounding the city, and our movements had drawn their attention. They were determined to disrupt the aid effort with heavier strikes on the approach roads.

I asked Max to fetch the first aid kit and body bag, but before he could, two border guards blocked him, gesturing firmly and shaking their heads. The situation unravelled quickly into raised voices. The guards insisted Borysko would be buried on site with the others who had fallen defending Bakhmut. To them, the priority was clear: evacuate the living, leave the dead.

Max bristled, shoving one of the guards back as Emma cried out in protest. Before it could spiral further, I forced myself to stay calm. Heart pounding, I walked to the ambulance and pulled out old embassy paperwork from my first visit the year before. Returning, I held it out and shouted a single word: "Diplomatico!", hoping it would work its magic once more.

Everything stopped. A senior officer emerged from the bunker, snatched the papers from my hands, and studied them under a small torch. The silence was unbearable. He finally raised his eyes, shining the light into my face before folding the papers with deliberate care. Handing them back, he surprised us by speaking in broken English: he was sorry for Borysko's death, grateful for our support, and understood our wish to take him home. But the responsibility would be ours.

Relief washed through us, mixed with dread. We shook hands, thanked him, and turned to the task ahead. Max and I would have to lead, and I accepted full responsibility for moving his body and taking it back to Kyiv.

We gathered around the body in reverence. I asked the others to recite the Lord's Prayer, then told Lucus and Emma to step away while Max and I dealt with the body. To my surprise, Lucus refused point blank to walk away, insisting that Borysko had been his friend and that he wanted to help care for him. Thinking back to the incident in Bakhmut with the rats, my mind turned inward to a darker place, making me pause before we could move forward. Together, we lifted the tarpaulin, revealing a mangled body with a shredded torso, a grey lifeless pile of twisted bones and flesh, all that remained of a young man's life, barely thirty, ended in an instant. His body was grey and cold to the touch, and unlike the bodies in AZOM, I found a profound sense of calm wash over me, founded on respect for another team member. There was no revulsion; it had been replaced with love for another human. I knelt slowly and found myself whispering an apology, begging his forgiveness. Though I had barely known him, he had been one of ours. The guilt struck deeply, "it should have been me, not him".

We unfolded the reinforced body bag and, as gently as we could, transferred him along with his belongings and what pieces of him we could recover. Rigor mortis made it difficult, but at least he was clean, untouched by infestation, unlike so many we had seen in Bakhmut. The guards stood silently, helmets removed, hands crossed, whispering prayers in their own tongue.

When we sealed the bag and placed him on the stretcher, Emma was waiting by the ambulance doors. Together we lifted him inside, securing him as carefully as any of the wounded.

We thanked the guards once more. Their faces had softened, sombre now, wishing us a safe journey back to Kyiv. For them, this was routine; for us, it marked the shadow that now hung over what should have been a successful mission.

We had planned to fill the ambulance with the wounded. Now, we made room for one more. Borysko was coming home.

CHAPTER 16

The Forest Camp

All of the others, along with many of the refugees, were waiting for us to return before they settled down for the night at the forest camp. By now, the facility had been overrun with people waiting to be moved, and we were acutely aware that the camp had been built in a very temporary manner from small wooden huts, tents, and makeshift shelters designed for a smaller group.

Luckily for us it was early spring, and the night remained above freezing. Having first visited Ukraine during the winter, I already knew how brutal the conditions could be, nearly freezing to death on my first night in Kyiv. Yet the Ukrainians, amongst many things, were resourceful. They had begun manufacturing large kitchen trailers capable of cooking a combined meat and vegetable stew in sixty-litre drums. They had been kind enough to keep some for us, knowing we were on our way.

It was a huge relief to arrive safely at the camp after such a sombre ride back from the checkpoint, with Borysko's body lying in state in the back of the ambulance. Emma and Lucus remained transfixed by the body bag. Even though we had lost him, the surviving members of the team were thankful that we had insisted he would come home with us. Many began to ask questions about where we had been and what we had seen, recognising from our clothes and dishevelled appearance that we had struggled to cope with the last week. It was then I realised I hadn't shaved properly in days and probably looked like an old, grey-haired hobo in body armour. We were tired and hungry, grateful simply for a hot meal. Sitting quietly, we reflected on what had happened and thought about the next two days ahead. Both Max and I struggled to answer their questions. It was too painful to describe the conditions in the trenches and the AZOM industrial area, and we needed to concentrate on what lay ahead.

For the first time in days, we could feel relaxed and reasonably safe. We were out of artillery range, in the amber zone, and on our way home. Just before we retired for the night, we requested a quick debrief and were pleasantly surprised to learn that the others had taken the initiative to begin evacuating the seriously wounded. Two of the larger lorries had already left for Kyiv the day before, acting on the advice of the camp commander who had been coordinating medical support. Even though this had been written into the initial plan, I had doubted it would happen without Max or me. But the situation demanded action, and the leaders of Owl, Hawk, and Black Swan had taken charge. Both Max and I congratulated them for their initiative in our absence.

For now, all we told them was that we had been sidetracked by the battalion commander, taken on patrol to view the tactical situation, and then trapped briefly by heavy artillery before returning with the last of the seriously wounded. After what had happened to Borysko, they were content with those answers, relieved simply that we had made it back. What became apparent, however, was just how lucky we had been. Several members of the team had suffered injuries from shrapnel, artillery, and ballistic impacts, and most of the vehicles had sustained some form of damage. I now understood the plan to use lighter, more manoeuvrable vehicles within the last four kilometres. Like the ambulance, many had been damaged, but they had absorbed the worst of the impacts, even from explosive drones or heavier shells. This now left us with functional lorries and larger vehicles in the forest to move the bulk of the refugees and wounded safely to Kyiv early in the morning.

The camp officer agreed to roster a watch overnight and understood the need for everyone to stay away from the ambulance, which had been locked. It had become painfully clear to me that Emma had grown fond of Borysko, as she wanted to remain near him. We therefore set up camp close to the vehicle, insisting she sleep with us as agreed from the start, always staying in groups of two or more.

Fatigue was etched into every face, streaked with dirt, dust, and mud-stained clothes. We were overcome with distress, unable to function further until rested. It was now midnight, and with plans to

set off before dawn, we were in desperate need of sleep. Finding whatever cover and comfort we could, we were asleep within minutes, collapsing wherever we could make ourselves even relatively comfortable.

Back on the Road

Max's stamina was amazing. He seemed able to survive on only three or four hours of sleep before moving again. By contrast, at my age I could feel the pain of old age creeping in. I struggled to come around in the morning until Max shook me into consciousness, handing me a large cup of coffee and a ration bar.

Over the past week, it had become painfully clear how important planning was, being self-sufficient, hiding enough rations, food, and supplies to keep ourselves going without becoming a burden on the locals. I had barely begun drinking my coffee while sharing out the food when I saw Marianne instructing a group of soldiers to load the first vehicle. It was barely twenty minutes before dawn and we needed to start moving. Like the initial convoy, all of the lorries would leave in staggered stages, led by the recovery vehicle, separated by ten minutes or more before closing the gap as we neared Kyiv. The system had worked like clockwork on the way here and was already swinging back into action.

Before I could finish shaving, I found myself standing before the team leaders, Max, Leo, and Lucus, gathered around me under the dim red glow of a head torch. After the usual pleasantries, I saw the look in their eyes and realised what they were about to say. We were already five people down and didn't have enough drivers to move the remaining lorries. When we had first approached Bakhmut, every vehicle had a driver and a fixer (translator). But with two vehicles already gone with Soren, the Swan group leader, and the loss of Borysko, we were now short one driver and two translators.

I knew Max well enough to sense his instinctive response, his notebook was already out. We suggested Emma drive the ambulance with one of the less seriously wounded soldiers acting as translator until we reached the warehouse. We shrugged at each other. It would have to do. We needed their help, and needed transport. Space was already scarce; we barely had room for the seriously wounded. Many of the men would have to stand in the trailers, gripping the rope netting we had fixed to the sides, or sit on hard plank benches. David and Sofia proposed asking a couple of the refugees, who spoke both English and Ukrainian, but Max and I had already agreed that civilians made poor fixers. They could too easily become argumentative at checkpoints, not out of defiance, but exhaustion, illness, or sheer desperation.

As I rinsed my face, scraping what felt like concrete dust out of my hair, I made a joke that its colour matched mine, "grey". Max chuckled and wandered off to find volunteers, while Leo and Lucus, coffee cups in hand, prepared the vehicles for loading. The key to this part of the mission was silence. Even though we were out of artillery range and under cover, the enemy could easily have spotted such a larger group on the outskirts and bombed us by air. We had to slip away quietly, scurrying into the darkness.

There was little left to say between us; we were all caught up in the work at hand. Having been one of the first into Bakhmut, I now felt I should be among the last to leave, riding with the ambulance and driving the final vehicle. Thankfully, all the lorries and vans started without issue. Engines roared to life, and once loaded, the convoy began its long return.

As the sun rose behind us, I thought about the others who had gone before. My body ached from head to toe. I was physically exhausted, promising myself a hot meal and shower once we reached Kyiv. It was going to be a long day, but the weather was good and the roads open. We were only twenty-four hours behind the first two vehicles, who had already briefed all the checkpoints. That made navigation easier. With our vehicles full of wounded troops, the guards often waved us through with sympathy.

The mood within the group was buoyant, which helped lift my spirits and push me forward as the day dragged on. Yet, in the back of my mind, there was Borysko, and his family. How would we tell them of his death? The closer we drew to the warehouse, the more it weighed on me, and I knew Max would be thinking the same. He had recruited Borysko locally through the legion.

Trying to think ahead to our arrival, I knew the warehouse would be chaos, wounded to unload, ambulances to direct, hospitals and care facilities to coordinate. Still, I couldn't shake the thought of Borysko's family. During one of our fuel stops, I messaged Max. I suggested we ask a member of his church, and perhaps someone from Borysko's family, to meet us later at the warehouse after the main unloading. It was all I could think about.

Solemn Duty

By the time we reached the warehouse on the outskirts of Kyiv, it was overflowing with people waiting for our convoy. The mood was almost triumphant. Families pressed forward, searching the vehicles for loved ones, relief and joy breaking through the weariness. Many of the wounded were rushed away at once by medical staff or carried into waiting ambulances.

As I watched the scene unfold, I felt a flicker of concern that the handover was too chaotic, but then I noticed the designated team leaders with their clipboards, carefully checking off names as each patient disembarked. Order, even in this storm.

Max was sitting on the edge of the loading dock, his shoulders heavy, eyes far away. Emma and I walked over to him. We were done, spent, ready for rest and a good wash like everyone else. The only task remaining was the final and most difficult one: handing over Borysko's remains to his family.

Despite the weight of that duty, Max and I allowed ourselves a brief moment of pride. We had managed a huge aid run, bringing out dozens of wounded from a city that was collapsing by the hour. Bakhmut would not hold much longer, but at least we had given many another chance. The losses still gnawed at me. Leaving Ned and Ella behind troubled me deeply, but I forced myself to remember, it had been their choice.

Max broke my thoughts by saying he had contacted a priest from St. Sophia's Cathedral, who was on his way with a member of Borysko's family. In the meantime, with most of the team still present, we gathered them for a short debrief. I congratulated them on their courage under impossible conditions. They had faced danger, exhaustion, and loss, yet carried out their duties with remarkable resolve. I told them I intended to write a book one day, naming every volunteer who had stood with Ukraine. They laughed and shook their heads. Not one of them wanted to be called a hero. In their eyes, they had only done their duty, nothing compared to those who had remained behind in Bakhmut. I could only respect that humility and understood it completely.

As the group dispersed, a car pulled up. Out stepped a priest in black robes, accompanied by an older man. My heart sank. Max exchanged a glance with me, then Emma quietly asked to stay with us.

Max greeted them first, speaking softly in Ukrainian. I didn't catch the words, but I understood the tone, the weight of loss shared between them. He introduced us: Father Shevchenko, associated with St. Sophia's, and Kolyma, Borysko's uncle.

Kolyma shook each of our hands firmly, then embraced us one by one. His voice broke as he thanked us for bringing his nephew home. I swallowed hard against my own rising grief, managing only to say that it was the least we could do. Father Shevchenko placed a reassuring hand on my arm, telling us that the church would care for Borysko and his family, that he would be remembered among the martyrs of Ukraine.

We walked together toward the ambulance. At the rear doors, Kolyma paused, his hand resting on the metal. He asked about the shell damage across the side panels, whether it had been sustained when Borysko died. When Max explained it had happened earlier, on one of our runs, Kolyma exhaled sharply and embraced us again. He understood, now, that we had shared the same dangers as his nephew.

Arrangements had been made to take Borysko to a church where his family had gathered. From there, his body would go to the mortuary, where the legal documents could be completed. Strictly speaking, we had bypassed Ukrainian procedures by bringing him directly from Bakhmut, but the family would untangle the bureaucracy later. Max quietly pulled me aside and asked whether I would be willing to state that Borysko had died on the journey, rather than in Bakhmut itself. It would make the paperwork simpler, spare the family extra hardship. I agreed without hesitation.

After a blessing at the ambulance, we lifted the stretcher down. Together, Max, Emma, Kolyma, and I carried him into the waiting vehicle.

The church stood with its doors wide open, candles glowing inside, incense drifting out into the cool air. A small crowd of mourners had gathered, but I saw only one face: Borysko's mother. Her eyes were swollen, her hands trembling as she reached toward the stretcher. Kolyma stepped forward, gently touching her arm, though he said nothing. There were no words.

We carried Borysko slowly up the steps, each creak of the boards heavy with meaning, and laid him before the altar on a prepared table. Max murmured a prayer under his breath, his voice low and steady.

I remained silent, unable to trust my voice.

His mother dropped to her knees, clutching the body bag as though holding on could keep him here. Her sobs filled the nave, echoing off the stone walls, raw and piercing. Kolyma knelt beside her, his hand resting gently on her back. The rest of us knelt too, sharing the moment, carrying the weight together.

Time seemed to pause. The war, the convoy, the chaos of the city had faded away. What remained was love, grief, and the solemn duty of bringing a friend home.

Borysko was no longer just another casualty of war. He was a son, a comrade, a brother. And now, finally, he was at peace.

CHAPTER 17

The Wake

The first thing we did on returning to the hostel was to unpack and jump in the shower before bed. While we had normally been limited to four minutes of hot water I couldn't help standing still under the water hoping that it wash away all of my sins together with the concrete dust that had now been matted into my hair for over a week.

On returning to the hostel we had all been immensely surprised and deeply touched to find people who had stayed up late knowing that we were on our way back jumping up to help us unload the ambulance instantly seeing the damage to the vehicles side and our dishevelled appearance. Many of them knew where we had been and were members of local battalions, people from around the world, anywhere from Europe, America and even Asia, all of them fighting in some way to support Ukraine resist Russia. They were understandably keen to hear our story and find out what had happen but none of us had the capacity to relive the events of the last days even though it had been very successful overall. We also knew that there were also members of the press at the hostel, we had already agreed not to say anything until Max and I could talk to his battalion liaison officer to compile an official statement, even in our circumstance there was a courtesy to be followed.

By the time I woke the next morning and gone down to put together breakfast, Max, Emma, and most of the team were gathering in the small basement kitchen sharing what we had stored in the fridge before we had left. Others now stood in the hallway outside offering us cups of hot coffee which they had made or bought for us, the strength of human compassion was overwhelming and had always been my weak point. After years of military training, I had discovered that I had almost always been able to keep myself together and composed during a fire fight, air raid or critical incident but suffered afterword when the dust had settled when asked to relive the moments again. And I was going to struggle through the next couple of days before leaving to go home.

Fortunately, most of those asking us questions understood that there would be an official statement due shortly and would need to wait for news, those who had decided to be more persistent had been politely asked to leave.

All of the team knew that I would be driving back to the border with Max the following day to disappear into the mists of time and kept the knowledge a secret. I just needed a day to recover, wash my clothes, sort out any outstanding paperwork and arrange for a team debrief that evening in the bar to honour Borysko. On the way down,

I had already caught the hostel manager asking him for a quiet corner of the lounge bar for the team to which he had kindly agreed to arrange with a free drink for all involved. He had known Borysko and asked to join the wake.

After breakfast a couple of the team and I took a final walk from the hostel to St Sophia Square again only ten minutes away to take a final look at the collection of burnt out armour and take a break within the sanctity of the cathedral. There was such a stark difference of atmosphere aside the cathedral compared to the chaos of the past week in Bakhmut, that I found myself quietly sobbing in silence praying for a better world asking America and Europe for more support for Ukraine. The suffering of the people was overwhelming. Before I knew it a priest had come up behind me offering solace. Unable to understand him, I just apologised and had been surprised when he replied with a good understanding of English. We sat together for hours as we discussed the past two years that I had experienced in Ukraine and the problems that had developed at home. Like any good Chaplain he listened and thanked the team and everyone involved for their support. He then asked me for my own Chaplain credentials before we shook hands parting.

I spent the rest of the day laying quietly on my bunk back at the hostel waiting for the evening trying to come up with something inspiring and fulfilling to say to the team. Max arrived knocking at the door a couple of hours before the wake and my last night. It was good to see him and quietly chat about he past years that we had grown to known each other remembering how cold I had been when we had first met during the autumn of 2022 just after the full scale invasion, and how Max had shared his breakfast with me on that first day. He was a good man and for the first time in years, I began to understand why I had stayed the first time and more importantly why I had come back, it had been for Max and people like Finn and Andrew from the Irish Medical Mission, genuine people who just wanted to live in a better world free of repression able to make their own choices without being dictated to by the authorises. We were free and willing to fight for it, and even die for it, that is what I was going to highlight to the others when remembering Borysko. We all shared the same values even though many of us came from different areas, backgrounds and cultures from all over the world. We were all willing to stand up for a free world and justice within our own societies, something that I still had to face at home.

The wake for Borysko was held in the lounge bar which had now been cordoned off by the hostel owner in honour of our friend, a modest room lit with warm lamps and filled with the hum of voices that carried both grief and gratitude. Friends, family, and comrades gathered around tables laden with simple food and beer, sharing stories in low tones. Despite the sorrow, there was an unmistakable sense of pride in the air, a collective recognition that his life had been lived with purpose and integrity.

One after another, people rose to speak of the man they had known:

a young man who never failed to visit his uncle on Sundays, a friend who would give the shirt from his back, a volunteer who carried wounded strangers as though they were family. His humour, his generosity, and his quiet courage painted a portrait of a man both ordinary and extraordinary.

As I listened, I realised how little time I had truly spent with him, how much more there was to know. I had worked alongside him, trusted him in dangerous moments, and yet I envied those who had known him for years. They spoke of his stubborn kindness, of his unshakeable faith, of the way he could make even the darkest day bearable with a simple smile.

When it came time for me to raise a glass, I could only say this: I wish I had known him better. But even in the short time I did, I knew he was genuine, and good, and that the world would be a smaller place without him.

As the gathering came to a close, I asked all of the team to stay behind for one last group huddle, thanking them for the courage, friendship and professionalism. The aid run to Bakhmut would have been a failure without them and I would always stay in touch even though I lived in the most isolated city on the planet, Perth, Australia.

Farewell Brother

There was little more I could say to Max except thank him for driving me back to the border crossing near Kraków.

At first, I had only expected him to take me as far as the border, an eleven-hour journey along the E40 motorway that would consume most of the day. He understood why I needed to avoid public transport, especially Lviv, after the incident at Smart Medicinal Aid. More than anyone, like Finn and Andrew, Max understood how quickly situations could unravel in Ukraine, especially in war-time with organised crime feeding on the chaos. My plan had been simple: reach a quiet border, cross on foot, catch a bus to the train station, then travel on to Kraków and Warsaw before flying home.

I should not have been surprised when Max insisted on taking me all the way to Kraków, retracing a route I had crossed once before. By the time we reached Brozhky on the M10 it was already too late to cross. We spent the night in the ambulance, reasonably comfortable on the stretcher and floor. I insisted Max take the stretcher, though we had shared the driving. The only concern nagging me was the ambulance's condition, one window still boarded up. Max's solution was simple: he would tell the guards he was taking it into Poland for repairs. Plausible, and a good idea.

Dawn broke with birdsong, a startling contrast to the past weeks of air raids, artillery, and gunfire. The peace felt almost foreign. As we ate breakfast in the sunshine, I began to see why Max had wanted to come this far: so he, too, could pause, repair the ambulance with friends, and breathe again. In that stillness we spoke of my first visit to Ukraine, of Rava-Rus'ka just north of where we were. For a moment we remembered a Ukraine at peace, and what it could be again.

The final goodbye came outside the Ibis Hotel near Kraków's main station. As I unpacked my bags, Max broke the tension before it could take hold with a joke about my poor Polish, warning me to buy the correct ticket, otherwise I would end up back to Lviv instead of Warsaw. He wasn't wrong; my vocabulary stretched only as far as the translator on my phone. The distraction worked, making parting easier. We exchanged military patches, mine from Australia, his from the Ukrainian Legion, and promised to stay in touch. I walked away before emotion overcame us.

On the train to Warsaw I laughed quietly, having bought the wrong ticket and needing to purchase another when the ticket inspector came by. Max had predicted it. For the first time in years, though, I felt different. Friends and Michelle, my wife, had noticed the change already. Full mechanised warfare alters a man, even an aid worker forced to endure the front-lines.

I longed to see Michelle again, knowing I would never return to Ukraine, though I would support them however I could. One more night in Warsaw lay ahead: a hot bath, a steak, and a cold beer, though I rarely drank.

Yet visions clung to me, The Bucha massacre in 2022, Kherson's liberation, and now Bakhmut's imminent fall. I thought of Ned and Ella still there, and the thought chilled me. Home would bring its own battles: Damo's mess, the looming court case, and the struggle to readjust. I knew I would need help to keep the monsters at bay.

But for now, I was simply grateful. Grateful to rest, to sleep on the flight home, and to carry the knowledge that, somehow, I had made it back.

CHAPTER 18

Australia

I had only a couple of weeks before my case was due for review, and I desperately needed a competent lawyer. The first I had engaged in Bedford, Perth had been disastrous, bleeding my family of funds during the discovery phase, delaying hearings to inflate costs, and failing even to provide proper invoices. The time in Ukraine had been bad enough, but facing trial for something I had not done was unbearable. As the date approached, though, support began to grow. The prosecution had no evidence, how could they? It was Damo who had stolen the money, processing it through his own accounts. We even had evidence of his alleged threats against me, passed on through Ned, who I prayed would survive Bakhmut with Ella.

The more we examined the prosecution's case, the more apparent it became that something was again deeply wrong. The police evidence was entirely circumstantial; their chain of custody broken, their witness statements conflicting.

It was beginning to look like a simple miscarriage of justice.

During this time, I kept working part-time, coordinating aid into Ukraine, sometimes covertly into Kyiv through UK Aid, Ukraine Patriots, and UK Med. At home, I supported Australian veterans with PTSD and men with disabilities through the Men's Shed. But the looming trial weighed on me and the family daily eroding our health.

When the trial finally came, the first lawyer failed to appear, ignoring a summons and risking his own reputation. By sheer luck, we found a new lawyer on the very first day, almost by chance outside the courthouse in Joondalup. From the moment Michelle and I sat down with him, relief washed over us. His background as both a police officer and now a criminal defence lawyer gave him unique insight. Within minutes, he identified flaws in the case and remarked on the police's failure even to investigate Damo properly.

Alex was a towering figure, well over six feet, broad-shouldered, immaculately dressed. His presence filled every room he entered. Charismatic and commanding, he carried himself with a confidence earned from years of experience. Judges respected him, juries trusted him, and even his opponents grudgingly admired his skill. His deep, measured voice carried authority, and his ability to weave complex arguments into simple, persuasive truths was remarkable.

Despite his size, he had a warmth in private. He listened carefully, asked sharp questions, and projected calm certainty. His confidence was infectious; with him by our side, I began to believe justice might prevail.

In court, Alex was relentless but precise. Over two days, he dismantled the prosecution piece by piece. Witnesses contradicted each other. The senior police officer shifted blame to his junior, who in turn blamed him back. Neither could explain why I had been charged instead of treated as the victim. Their case unravelled to the total dismay of the magistrate.

When the time came for me to give evidence, there was no need. The prosecution's case had collapsed entirely. All that remained was to trust the magistrate's judgement, due to be delivered in the new year.

End Game, Justice

Early in the new year of 2025, the judgement in my case was delivered. Michelle, my lawyer, and I were astounded by the detail with which the magistrate meticulously walked through the evidence before issuing a "full acquittal". In the process, she openly condemned the serious misconduct of the senior police officers involved in the investigation.

The police had misrepresented crucial DNA evidence in a clear attempt to manipulate the case in their favour. The judge not only cleared me of all charges but went further, highlighting that the officers had been "untruthful." The irony was unmistakable: I had been charged with "creating a false belief," while in reality, it was the police themselves who had lied.

It took time for the consequences of the judgement to sink in. My community and colleagues in the Australian Federation of Ukrainian Organisations (AFUO) welcomed the outcome, but many remained troubled. The authorities had overlooked important evidence about Ukraine, regarding Smart Medicinal Aid, Damo's alleged fraud through "Funds for Community Aid Ltd", and the misconduct of the West Australian police.

With experience, I have come to understand that the most ethical decisions in life are often the hardest to make. Many authorities and institutions are incapable of processing the truth, too damaging, too uncomfortable. But if you know the truth in your heart, you must fight injustice, whether at home in your community or abroad in a war zone for the benefit of the people.

Armed with the transcripts and outcome of my case, the matter has now been referred to the Corruption and Crime Commission (CCC), Western Australia's peak anti-corruption body. They have been instructed to investigate the officers who misrepresented evidence, with calls for a full, independent review into WAPOL's (The West Australian Police), handling of the case, covering misconduct, evidence tampering, and abuse of process.

In addition, there are calls for a federal review of NGO operations abroad, particularly in Ukraine, to ensure transparency and accountability after the disappearance of donated funds linked to Funds for Community Aid Ltd. Most importantly, the case has underlined the urgent need for stronger protections for International Aid workers, and (ANZAC), Australian and New Zealand civilian volunteers working in conflict zones.

As far as Funds for Community Aid Ltd are concerned, and in light of Damo's conduct, he has become a casualty of the war, unable to cope with the distress of experiencing the war, up close and personal. You may ask why I care. The simple truth is that I introduced Damo to Ukraine Patriots and therefore feel partly responsible for his actions. Processing donated funds from private clubs and the public through personal accounts, rather than the designated company account, could reasonably be viewed as alleged fraud. The West Australian legal team has already submitted full evidence packs regarding his conduct to both the WA and NSW Police Commissioners. Yet each jurisdiction attempted to defer responsibility to the other, leaving the matter unresolved. While I can forgive him as a man, I cannot accept the harm he caused to others.

The authorities are aware, but the embarrassment caused by their failed prosecution of me, the exposure of their own misconduct has weakened their willingness to pursue him. In the meantime, Damo has been discredited by his failure to provide complete financial accounts for Funds for Community Aid Ltd company, despite repeated requests. The aid community, in both Australia and Ukraine, know the truth and steadily begun to distance themselves from him.

The third problem with Smart Medicinal Aid has only deepened since my assault in 2022. Irena, the CEO, now struggles to attract large-scale overseas donations as increasing first-hand accounts of her alleged embezzlement come to light. Too many witnesses have spoken of her selling donated aid for personal gain. In time, justice will prevail there too.

As in Australia, the Ukrainian police are overstretched and under resourced, with many of their staff conscripted into the military, leaving them unable to deal with allegations of corruption in civil society. High-profile institutions, such as the Catholic College, further complicate matters when implicated.

Ukraine is not fighting in isolation; it stands on the forward edge of a wider contest over sovereignty, democratic order, and the rule of law in Europe. The country has endured immense losses, human, economic, and cultural, while resisting sustained aggression from Russia. In doing so, Ukraine has effectively become the shield for a broader international system that depends on territorial integrity and collective security.

Yet weapons and humanitarian aid, though essential, are not sufficient on their own. Sustained victory also depends upon institutional strength and moral credibility. Wartime conditions inevitably strain governance structures and create opportunities for exploitation. Organised crime networks and fraudulent organisations can thrive in environments where urgency eclipses oversight. For Ukraine to prevail not only militarily but politically, it requires continued international partnership in strengthening anti-corruption frameworks, enhancing financial transparency, and protecting legitimate civil society actors.

Exposing fraudulent NGOs and criminal enterprises is not an act of criticism against Ukraine; rather, it is an affirmation of its long-term resilience. Transparent oversight mechanisms, cooperative intelligence sharing, and principled donor engagement reinforce public trust and safeguard the integrity of humanitarian operations.

In modern conflict, legitimacy is as strategic as artillery. If Ukraine is bearing the cost of defending a rules-based order, then the international community carries a parallel responsibility: to ensure that support, financial, logistical, and moral, is delivered with accountability, integrity, and unwavering commitment to the values being defended.

Working in the Aid Community

Working within the aid community, whether locally or abroad, requires a high degree of flexibility and cultural sensitivity. It means being prepared to adapt to unfamiliar environments, respecting the traditions and customs of the communities you are serving, and acknowledging that in times of conflict, corruption or "commission fees" can become part of the operating environment. At the same time, there is a moral responsibility to recognise when these practices cross the line, when they are no longer a reflection of local norms but instead constitute exploitation or outright crime.

This distinction became painfully clear in the case of Smart Medicinal Aid in Lviv. What initially appeared to be standard local practice was soon exposed as large-scale illicit trade, with hundreds of thousands of Euros and/or Dollars being siphoned away. Andrew and the firsthand witness from Lifeline Ambulances were the first to uncover tangible evidence of these activities, and my own arrival only compounded the situation by bringing further truth to light. At that point, we had no choice but to report what had been discovered, despite knowing the risks. Taking such action is never easy; it places you in direct conflict with individuals and organisations who may retaliate, and it takes a toll on both your physical safety and your emotional well-being. This is why resilience is not optional in humanitarian work, it is essential.

After three years of sustained involvement in Ukraine, working on the ground, remotely coordinating aid deliveries, and contributing to capacity building, I now feel a profound sense of relief that justice has prevailed. For the first time, I have been able to step back, rest, and begin to recover from the relentless demands of this work.

Among the many moments of recognition during this period, one of the most meaningful for me was receiving a small plaque, an international aid award from Father Shevchenko in association with St Sophia Church in 2024. This honour was presented in acknowledgement of the recovery of Borysko and the wider humanitarian efforts in which I was involved in. For me, this simple award symbolises not just the outcome of years of hard work, but also the resilience and faith that has carried both myself and many others through the darkest of times.

In gratitude for the inspiration and encouragement I have received from Chaplaincy Australia, and in enduring faith that Ukraine will prevail, I have passed this award on to them for display. My hope is that it will serve as a small but powerful reminder to others of the importance of reaching out, offering support, and standing alongside those who fight for freedom and humanity.

Drones

As far as drone technology is concerned in the ongoing conflict in Ukraine, new off-the-shelf and composite drone technologies have emerged as a game-changer, altering the dynamics of modern warfare in a profound way. Anyone can buy a commercially available drone and use it in the fight against Russia.

As both Ukrainian forces and separatist groups adapt to these technologies, they have significantly impacted the strategies, tactics, and outcomes of war. This book explores the multifaceted impact of drones on the Ukraine conflict, shedding light on their role as a powerful tool of war in the 21st century.

One of the most prominent roles of drones in the Ukraine conflict is their extensive use for aerial reconnaissance. Equipped with advanced cameras and sensors, drones provide real-time intelligence to both sides of the conflict. They offer a cost-effective means of monitoring

enemy movements, identifying troop deployments, and assessing battlefield conditions. This level of situational awareness has become invaluable for military commanders, allowing them to make informed decisions in real-time. Armed drones have taken precision warfare to a new level in Ukraine. They enable precise and targeted strikes against enemy positions, reducing the risk of collateral damage compared to traditional artillery or airstrikes. This precision has become crucial, especially in urban warfare scenarios, where civilian populations are often in close proximity to military targets. The ability to hit specific enemy positions while minimising harm to non-combatants has reshaped the ethical considerations of warfare.

Mobile Drone Unit, in the field Recon

Beyond physical damage, drones have a significant psychological impact on the battlefield. The constant threat of drone attacks creates fear and uncertainty among soldiers and civilians alike. Drones disrupt enemy morale and force troops to remain on high alert, impacting their overall combat effectiveness. This psychological warfare aspect has made drones a potent tool for asymmetrical warfare, where insurgents can challenge conventional forces.

Drones have also been employed to target supply convoys,

disrupting the flow of essential resources to both sides of the conflict. By targeting logistical operations, drones can strain the sustainability of military operations, forcing combatants to adapt and develop more resilient supply lines. The disruption of supply chains adds a new dimension to the conflict's strategic calculus.

DJI Drone, modified to carry, Explosive Ordnance

Drone defence, as drones have proliferated in the Ukraine conflict, so too have efforts to counter them. Anti-drone technologies, including jamming systems and anti-aircraft weaponry specifically designed to neutralise enemy drones, have seen increased use. This cat-and mouse game between drone operators and anti-drone measures has further complicated the battlefield, requiring constant adaptation and innovation on both sides. International Concerns, the widespread use of drones and AI (Artificial Intelligence) in the Ukraine conflict has raised international concerns about their proliferation and regulation in conflict zones. The ease with which drones can be acquired and operated has prompted discussions about the need for stricter controls. The potential for drones to be used in future conflicts and their implications for international law and ethics have become pressing topics in global warfare.

Custom made composite drone, Mission specific payload

The impact of new drone technologies on the Ukraine war is undeniable. From enhanced reconnaissance capabilities to precision strikes and psychological warfare, drones have reshaped the strategies and tactics of modern warfare. As the conflict continues to evolve, so too will the role of drones, highlighting the need for ongoing international dialogue on the regulation and ethical use of these powerful tools of war. Their influence in Ukraine serves as a stark reminder of the evolving nature of conflict in the 21st century and is a lesson for UK, America and Australia and their allies together with the rest of the world to learn from.

Using commercially available small drones like the DJI used in conjunction with AI, machine learning and swarm technologies will dominate modern warfare. Unlike the larger military, grade drones that cost hundreds of thousands, smaller, cost-effective drones available at a fraction of the price now dominate the battlefield at an infantry level. Warfare will never be the same.

Organised Crime

I have come to realise that organised crime and war are deeply intertwined. Conflict creates opportunity, and criminal networks are quick to exploit it. Under the guise of humanitarian aid agencies, some gangs move into war-torn countries, embedding themselves within local systems. They cultivate relationships with officials and community leaders, using them, knowingly or not, as cover for illicit operations. This regrettably makes it harder for genuine humanitarian organisations to function and earn trust.

Those who truly wish to support Ukraine must do so through direct, personal relationships with trusted local communities, people they know and have worked with before, whose integrity has been proven. I was fortunate to meet several genuine aid agencies and military volunteer groups, four or five in total, whose members risk their lives daily to bring real assistance to those in need. You can usually tell who is genuine: they are the ones working on the front lines, not sitting comfortably in offices in secure cities like Lviv.

There are, however, persistent allegations that certain organisations, such as Smart Medical Aid and others, may be involved in corruption. Those who have spoken publicly against them have, in many cases, vanished or been systematically discredited. Time will eventually reveal the truth, but for now, the public must make its own judgement based on limited evidence and witness testimony.

Ultimately, anyone working in a conflict zone must accept that some level of corruption in the aid sector is almost inevitable. The hard question is one of tolerance: if eighty percent of the aid reaches the people who truly need it, is that acceptable? Or is it simply complicity by another name? Each of us must decide whether we can live with that compromise, or risk everything to expose it.

Characters

Australian Ukraine Embassy

Volodymyr – Embassy's Chargé d'Affaires

Australia

Damo – CEO, Funds for Community Aid Ltd

Mike – Director, Funds for Community Aid Ltd

Kyiv

Dream Hostel Team

Max – DH Volunteer Team Coordinator and Ukraine Legion Member

Jen – journalist, writer and radio producer from Scotland.

Ned – Australian ADF veteran, medic Assoc and drone specialist

Ella – KiWi Nurse and Ukraine front line medic

Nastasiya – Kherson local and volunteer Fixer

Kolyma, Borysko's uncle

Father, священик (svyashchenyk) – Shevchenko

Lviv

Smart Medicinal Aid

Irena – CEO, Smart Medicinal Aid

Yermani – 2IC, Smart Medicinal Aid

Bruce – Mercenary One, Smart Medicinal Aid

Liam – Mercenary Two, Smart Medicinal Aid

Tetyana – Associate, Smart Medicinal Aid

Vasily – Ukraine Government Contact

Caitlin – Lifeline Ambulances Ireland

Kherson

Nikolai – Battalion chief drone operator/instructor

Finn House Films

Finn – Documentary Films and News Reporter

Andrew – Associate of Finn House Films,

       Ex Smart Medicinal Aid 2IC,

       Became CEO The Irish Medical Mission

Garik –   Covert Ukrainian Fixer

The Battalion Commander(s) –Classified

Mikhail Troop Sargent – 1st Corps, Ukrainian National Guard "AZOM"

Bakhmut Aid Mission

Call Signs

Nightingale (Соловей, Solovey)

Max – Nightingale One

Marianna - Nightingale Two

Peter Savage - Nightingale - Three

Owl (Сова, Sova)

Léo - Owl One

Aleksy - Owl Two

Hanna - Owl Three

Hawk (Яструб, Yastrub)

Lucus - Hawk One

Emma - Hawk Two

Borysko - Hawk Three

Swan (Лебідь, Lebid).

Soren - Swan One

Sofia - Swan Two

David - Swan Three

Epilogue

SLAVA UKRAINE"

"Боротьба від імені людства"

The Aid Agencies

Acknowledgements

UK Aid for Ukraine

A specialised Ukrainian based NGO supplying aid to the most vulnerable communities on the front lines.
www.facebook.com/groups/930637394275431

DIY Ukraine.
UK based NGO dedicated to supplying aid to Ukraine.
www.facebook.com/diyukrainegroup

Ukraine Patriots, Kyiv

A group of Ukrainians and internationals working tirelessly to aid

volunteers defending Ukraine and civilians caught in the crossfire.

Unbroken Lviv

Specialised rehabilitation and recovery

www.unbroken.org.ua

UK Med

UK-Med is a frontline medical aid charity.

Provides mobile life saving surgical services.

www.uk-med.org

The governments and people of the United States, UK and Australia who have played a critical role in sustaining the defence of democratic principles during this conflict.